HOW
SPACE
WORKS

HOW
SPACE
WORKS

CONTENTS

Editorial Consultants
Anthony Brown,
Dr. Jacqueline Mitton

US Editor
Kayla Dugger

Art Editor
Steve Woosnam-Savage

Project Art Editors
Amy Child, Shahid Mahmood,
Jessica Tapolcai

Illustrators
Mark Clifton,
Phil Gamble, Dan Crisp

Managing Art Editor
Michael Duffy

Jacket Designer
Tanya Mehrotra

Senior Production editor
Andy Hilliard

Art Director
Karen Self

Contributors
Abigail Beall, Philip Eales,
John Farndon, Giles Sparrow,
Colin Stuart

Senior Editor
Peter Frances

Project Editors
Nathan Joyce, Martyn Page

Editors
Annie Moss,
Hannah Westlake

Managing Editor
Angeles Gavira Guerrero

Senior Producer
Meskerem Berhane

Publisher
Liz Wheeler

Publishing Director
Jonathan Metcalf

First American Edition, 2021
Published in the United States by DK Publishing
1450 Broadway, Suite 801, New York, NY 10018

Copyright © 2021 Dorling Kindersley Limited
DK, a Division of Penguin Random House LLC
21 22 23 24 25 10 9 8 7 6 5 4 3 2 1
001–319132–Mar/2021

A catalog record for this book
is available from the Library of Congress.
ISBN 978-0-7440-2748-8

This book was made with Forest Stewardship
Council ™ certified paper—one small step
in DK's commitment to a sustainable future.
For more information go to
www.dk.com/our-green-pledge

Printed and bound in China

For the curious

www.dk.com

SPACE EXPLORATION

SPACE FROM EARTH

Earth

Venus

The Oort Cloud

The Moon

The Sun

Saturn

The Kuiper Belt

Nearest star (Proxima Centauri)

| DISTANCE FROM EARTH | 600,000 MILES | 60 MILLION MILES | 6 BILLION MILES | 6 X 10¹¹ MILES |

Earth's diameter is 7,930 miles (12,760 km); the Moon is 238,855 miles (384,400 km) away

The rocky inner planets lie within the Main Belt of asteroids, which is 2.5 times farther from the Sun than Earth

All the planets in the Solar System orbit our local star, the Sun

Beyond the planets is the Kuiper Belt, 9 billion miles (15 billion km) from the Sun

From Earth to the cosmic web
Everything in the Universe, from our planet to clusters of galaxies, is part of a structure. If we could zoom out on the Universe, we'd see an interconnected web of galaxies and gases, called the cosmic web.

The Solar System is part of the Milky Way Galaxy, which includes about 100–400 billion stars

EARTH AND MOON

THE INNER SOLAR SYSTEM

THE SOLAR SYSTEM

The disk of the Milky Way is 100,000–12,000 light-years across

THE MILKY WAY

Structures in the Universe

Everything made of matter in the Universe—including the densest stars, planets, and moons, as well as diffuse gas and dust—can be arranged in a hierarchy of structures, all bound together by gravity. Objects within a structure orbit a center of mass, usually in the center of the structure. For example, the planets in the Solar System orbit the central Sun, while everything in our galaxy orbits its center, which contains a supermassive black hole around 4 million times the mass of the Sun.

Our place in the Universe

The Universe is everything that exists, has existed, or will exist. It comprises all matter and all space, permeated with light and other kinds of radiation. It also includes all time, both past and future.

WHAT SHAPE IS THE UNIVERSE?

Because the Universe does not have a recognizable edge, we cannot say what shape it has. Some cosmological studies suggest it is flat, while other data indicates it might actually be round like a sphere.

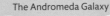

1,000 light-year sphere

The Andromeda Galaxy

Edge of observable Universe

...ere contains 90 percent of naked-eye stars

Center of the Milky Way

The Virgo Cluster

Nearest quasar

6 X 10¹⁵ MILES   **6 X 10¹⁷ MILES**   **6 X 10¹⁹ MILES**   **6 X 10²¹ MILES**

Cosmic distances
Distances in the Universe cannot be represented with a simple linear scale. On this chart, each division represents a distance 10 times greater than the previous division.

THE AGE OF THE UNIVERSE IS 13.8 BILLION YEARS

Size and distance

Outside the Solar System, distances become so vast that new units are needed to measure them. One of these units is the light-year, the distance that photons—particles of light or other electromagnetic radiation—cover in one year. A light-year is about 5.9 trillion miles (9.5 trillion km). The part of the Universe we can see, called the observable Universe, is limited by this distance, because light has had only the time since the Big Bang to reach us. We cannot see anything beyond this limit, known as the cosmic light horizon. The size of the whole Universe is unknown. One possibility is that it is infinite, meaning it has no edge.

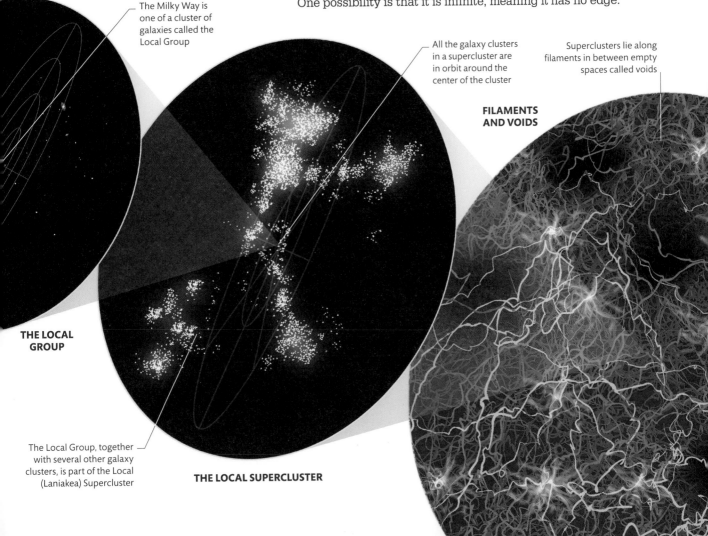

The Milky Way is one of a cluster of galaxies called the Local Group

All the galaxy clusters in a supercluster are in orbit around the center of the cluster

Superclusters lie along filaments in between empty spaces called voids

FILAMENTS AND VOIDS

THE LOCAL GROUP

The Local Group, together with several other galaxy clusters, is part of the Local (Laniakea) Supercluster

THE LOCAL SUPERCLUSTER

Looking into space

For most of human history, the Sun was thought to orbit Earth because of the way it moves in the sky. Now we know that Earth orbits the Sun and spins on an axis, too. Together, these motions create the apparent movement of the night sky around us.

The celestial sphere

The planets that are visible to the naked eye are much closer than the stars in the night sky. However, for the purposes of identifying the position of each celestial object, astronomers imagine everything, including stars, planets, and the Moon, as points on an imaginary sphere with an arbitrary radius around Earth. This is called the celestial sphere.

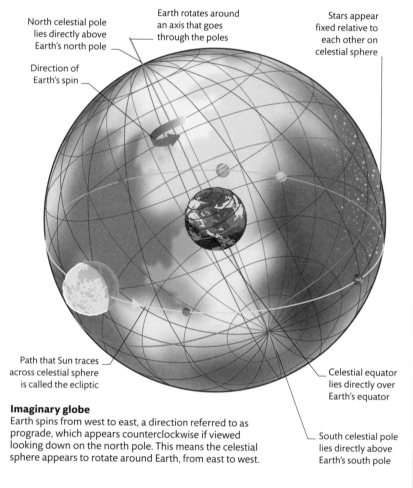

North celestial pole lies directly above Earth's north pole

Earth rotates around an axis that goes through the poles

Stars appear fixed relative to each other on celestial sphere

Direction of Earth's spin

Path that Sun traces across celestial sphere is called the ecliptic

Celestial equator lies directly over Earth's equator

South celestial pole lies directly above Earth's south pole

Imaginary globe
Earth spins from west to east, a direction referred to as prograde, which appears counterclockwise if viewed looking down on the north pole. This means the celestial sphere appears to rotate around Earth, from east to west.

How the sky changes

Over a day, the celestial sphere appears to rotate around Earth. This means that the stars, although fixed relative to each other, trace a circular path across the sky. Except for stars near the poles, most stars appear to rise and set. As Earth orbits the Sun, the stars that are visible at night vary through the year depending on Earth's position. This means that every night, the appearance of the night sky gradually shifts position. From one day to the next, if you were to look at the sky at exactly the same time, the stars would have shifted their positions by about 1 degree.

POSITION OF OBSERVER IN AUGUST

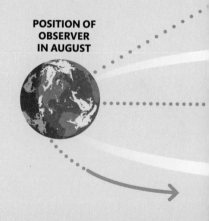

THE ECLIPTIC

Over a year, as Earth orbits the Sun, our star seems to trace a line across the celestial sphere. This path, the plane of Earth's orbit, is called the ecliptic. The other planets orbit more or less in the same plane as Earth and always appear near this line. The Moon orbits at a shallow angle in relation to the ecliptic, and eclipses only occur when the Moon travels through it.

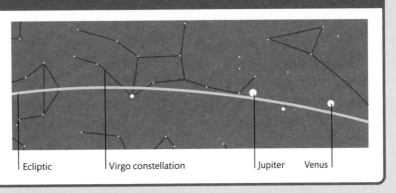

Ecliptic   Virgo constellation   Jupiter   Venus

Parallax

If you look at something with one eye and then through the other, it will appear to shift slightly. In the same way, objects in the sky appear in different positions depending on where Earth is in its orbit around the Sun. This is called parallax. The closer an object is to Earth, the farther it appears to move and the greater its parallax angle. This means that parallax measurements can be used to calculate distances to stars.

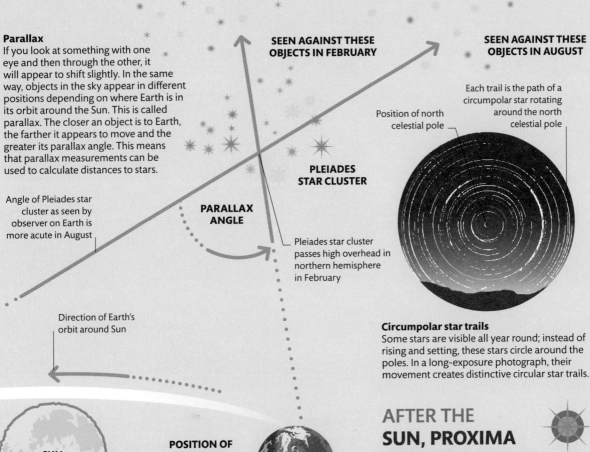

Angle of Pleiades star cluster as seen by observer on Earth is more acute in August

SEEN AGAINST THESE OBJECTS IN FEBRUARY

SEEN AGAINST THESE OBJECTS IN AUGUST

Position of north celestial pole

Each trail is the path of a circumpolar star rotating around the north celestial pole

PLEIADES STAR CLUSTER

PARALLAX ANGLE

Pleiades star cluster passes high overhead in northern hemisphere in February

Direction of Earth's orbit around Sun

Circumpolar star trails

Some stars are visible all year round; instead of rising and setting, these stars circle around the poles. In a long-exposure photograph, their movement creates distinctive circular star trails.

SUN

POSITION OF OBSERVER IN FEBRUARY

AFTER THE **SUN, PROXIMA CENTAURI** IS THE **CLOSEST STAR TO EARTH,** SITUATED APPROXIMATELY **4.22 LIGHT-YEARS AWAY**

Celestial cycles

To us on Earth, celestial events occur in cycles determined by the movements of Earth, the Sun, and the Moon. These cycles give rise to units of measurement for time, such as days and years, and to seasons. Related cycles are responsible for spectacular lunar and solar eclipses.

Why we have seasons

Earth orbits the Sun while spinning on its axis, which runs between the north and south poles. However, the axis of Earth's rotation is tilted about 23.5 degrees from the vertical in relation to the plane of the orbit around the Sun. This tilt means that there are certain points in its orbit where Earth's north pole is pointing toward the Sun and others where it points away. This tilt also means that the amount of sunlight Earth's north and south hemispheres receive changes over a year. The change in the amount of daylight in each hemisphere is the reason Earth experiences seasons.

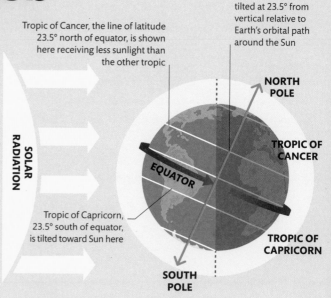

Tropic of Cancer, the line of latitude 23.5° north of equator, is shown here receiving less sunlight than the other tropic

Axis of rotation is tilted at 23.5° from vertical relative to Earth's orbital path around the Sun

SOLAR RADIATION

NORTH POLE

TROPIC OF CANCER

EQUATOR

Tropic of Capricorn, 23.5° south of equator, is tilted toward Sun here

TROPIC OF CAPRICORN

SOUTH POLE

The Earth's tilt
In the hemisphere that is tilted away from the Sun, solar radiation is spread out over a greater area of Earth's surface. This heats the surface less intensely, making it cooler than the other hemisphere.

Days and years

There are two ways of measuring days and years. A solar year, or a tropical year, is the time it takes Earth to return to the same angle with respect to the Sun. A sidereal year is measured using Earth's position relative to the fixed stars. The difference between the two is about 20 minutes. In the same way, a sidereal day is measured by Earth's rotation compared to the fixed stars, while a solar day is the time it takes for the Sun to return to the same position in the sky. The difference between the two is four minutes, because of the distance that Earth has moved in its orbit around the Sun during that time.

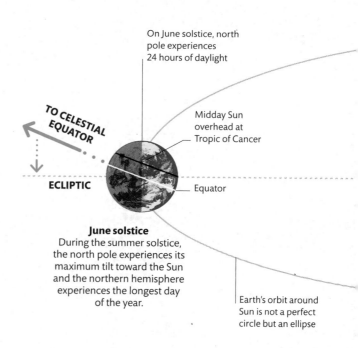

On June solstice, north pole experiences 24 hours of daylight

TO CELESTIAL EQUATOR

Midday Sun overhead at Tropic of Cancer

ECLIPTIC

Equator

June solstice
During the summer solstice, the north pole experiences its maximum tilt toward the Sun and the northern hemisphere experiences the longest day of the year.

Earth's orbit around Sun is not a perfect circle but an ellipse

Solstices and equinoxes
At the solstices, one hemisphere experiences its longest day, followed by the other hemisphere six months later. At the equinoxes, night and day are both exactly 12 hours long everywhere on Earth.

WHY DOES EARTH TILT?

Four billion years ago, when the planets of our Solar System were forming, Earth suffered a series of collisions with other planet-sized objects. The last of these, thought to have been with a Mars-sized planet, threw Earth's spin into a tilt.

EARTH IS CLOSEST TO THE SUN IN JANUARY, DURING THE NORTHERN-HEMISPHERE WINTER

PRECESSION

Due to gravity, Earth's axis of rotation is moving around, like a spinning top, in a cone-shaped motion called precession. It takes 25,772 years to complete one cycle of precession. This means that the north star, Polaris, will not always be situated almost directly above the north pole as it is now. Eventually, the star Vega will replace Polaris as the north star.

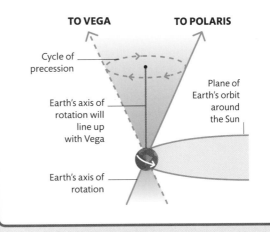

TO VEGA   TO POLARIS

Cycle of precession

Plane of Earth's orbit around the Sun

Earth's axis of rotation will line up with Vega

Earth's axis of rotation

Sun directly above equator at midday

December solstice
At the winter solstice, the north pole is at its maximum tilt away from the Sun. This means the northern hemisphere receives the smallest number of sunlight hours of the year.

TO NORTH CELESTIAL POLE

March equinox
On the spring equinox, Earth is tilted neither toward nor away from the Sun. On this day, the Sun sits exactly above the equator at midday.

SUN

Earth's axis tilts 23.5° from vertical

Midday Sun overhead at Tropic of Capricorn

Ecliptic—plane of Earth's orbit and Sun's path across celestial sphere

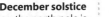

Earth spins west to east

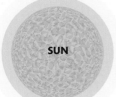

On December solstice, south pole experiences 24 hours of daylight

EARTH'S ORBITAL MOTION

September equinox
As with the spring equinox, Earth is tilted neither toward nor away from the Sun, and the midday Sun lies above the equator.

TO SOUTH CELESTIAL POLE

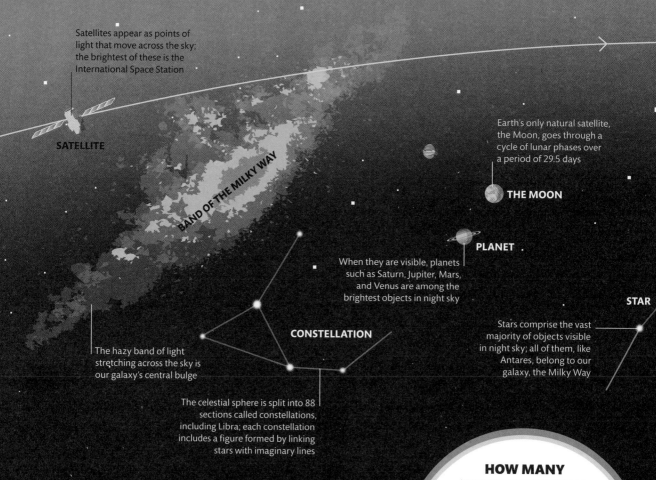

Satellites appear as points of light that move across the sky; the brightest of these is the International Space Station

SATELLITE

BAND OF THE MILKY WAY

Earth's only natural satellite, the Moon, goes through a cycle of lunar phases over a period of 29.5 days

THE MOON

PLANET

When they are visible, planets such as Saturn, Jupiter, Mars, and Venus are among the brightest objects in night sky

STAR

CONSTELLATION

Stars comprise the vast majority of objects visible in night sky; all of them, like Antares, belong to our galaxy, the Milky Way

The hazy band of light stretching across the sky is our galaxy's central bulge

The celestial sphere is split into 88 sections called constellations, including Libra; each constellation includes a figure formed by linking stars with imaginary lines

What can we see with the naked eye?

The night sky is an endless source of wonder, and it only requires a pair of eyes to see a variety of different objects. Over the course of just an hour on any given night, you will be able to see countless stars, at least one meteor, a satellite, and maybe even a planet or two. Away from light pollution, which makes the features of the night sky difficult to make out, the glow from our own Milky Way Galaxy's core of stars and dust shines like a faint stripe across the sky.

HOW MANY STARS ARE VISIBLE WITH THE NAKED EYE?

With perfect conditions and excellent eyesight, over 9,000 stars are visible to the naked eye, although at any given location, only half of these can be seen at once.

Objects in the sky

During the day, the light from the Sun dominates the sky so that nothing else, except the Moon, is visible. But at night, as we turn our backs on the Sun, the night sky reveals a wide variety of celestial objects, some visible to the naked eye and some revealed using magnification

Binoculars are needed
to see the Crab Nebula,
a remnant from an
exploding star

NEBULA

PLANETARY RINGS

The rings around Saturn
are only visible using
high-power binoculars
or a small telescope

METEOR

Meteors are bits of rock and
dust broken off from comets
and asteroids that enter the
atmosphere at high
speeds and burn up

Visible using binoculars and telescopes

Binoculars are portable and easy to use and
are a good way to see more objects and finer
detail in the night sky. Using the greater
magnification offered by a telescope opens up
even more of the night sky to an observer.

GALAXY

The Andromeda
Galaxy, 2.5 million
light-years away, is
the most distant
object visible with
the naked eye, but
it can be seen in
much greater detail
through a telescope

Visible with the naked eye

All of the celestial objects shown here
against the night sky are visible with the
naked eye on a clear night. The brightest
object by far is a full Moon.

What can we see with magnification?

There are plenty of amazing celestial
objects to see with the naked eye, but
equipment that magnifies these distant
objects reveals a new level of detail.
Through binoculars, the color of planets,
the details of nebulae, the craters on the
Moon's surface, and star clusters can all
be seen. With the smallest telescopes,
details like the rings around Saturn and
the shapes of nearby galaxies start to
appear. Bigger telescopes can peer
beyond our galaxy.

WHY STARS TWINKLE

Stars twinkle because of turbulence
in Earth's atmosphere. Changes in
density and temperature can cause
starlight to change direction
slightly. This effect is more visible
in stars than in planets, because
their light appears to come from
a single point, known as a point
source. It is also more prominent
in stars lower toward the horizon,
because the light has to travel
through more of the atmosphere.

Starlight has a shorter
path to travel, so star
will seem to twinkle less

Star appears to
twinkle more, as
its light has more
atmosphere to
pass through

NEARLY **EVERY STAR** YOU CAN SEE
WITH THE **NAKED EYE** IS **BIGGER**
AND **BRIGHTER** THAN **THE SUN**

Constellations

In astronomy, the night sky is split into sections called constellations. Historically, these were imaginary patterns of stars, but in the early 20th century, they were redefined as areas of sky. Although they might look like a group, the stars in a constellation are not necessarily close to each other in space.

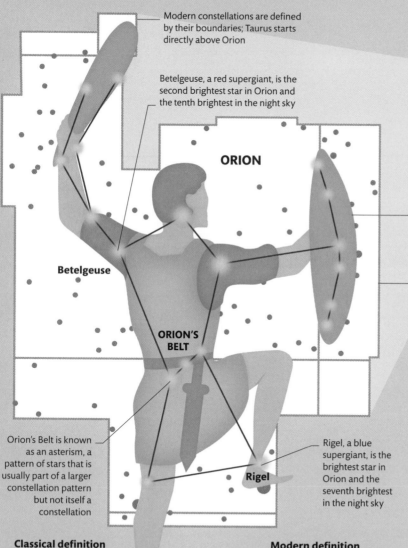

The 88 constellations interlock to fill the entire sky

CELESTIAL SPHERE

Modern constellations are defined by their boundaries; Taurus starts directly above Orion

Betelgeuse, a red supergiant, is the second brightest star in Orion and the tenth brightest in the night sky

ORION

Betelgeuse

ORION'S BELT

The pattern created by imaginary lines drawn between the stars resembles the classical figure of Orion

The boundaries of modern constellations are straight lines, either horizontal or vertical

Orion's Belt is known as an asterism, a pattern of stars that is usually part of a larger constellation pattern but not itself a constellation

Rigel, a blue supergiant, is the brightest star in Orion and the seventh brightest in the night sky

Rigel

Classical definition
Originally, the constellations were defined by patterns made from the stars. These were picked out because of their resemblance to animals or gods.

Modern definition
Instead of being defined by the stars that make up a pattern, the 88 modern constellations are shaped by their boundaries. These were drawn in 1928 and together cover the entire celestial sphere.

Patterns in the sky

Constellations are a way of grouping stars together. There are 88 official constellations recognized by the International Astronomical Union. These are often depicted by drawing lines between stars to mark out a pattern. However, constellations are actually defined by their boundaries, not by the patterns the stars create in the sky. Together, the 88 constellations cover the entire celestial sphere (see p.12). Every star that falls within a boundary is part of that constellation, even if it is not one of the main stars creating the pattern.

Band of the zodiac
The zodiacal band is part of the night sky in which the ecliptic, the planets, and the Moon are all found. It extends approximately 8° on either side of the ecliptic.

Ophiuchus, least known of zodiac constellations

Capricornus, the sea goat, is smallest constellation in zodiac

Aquarius is found in part of sky known as "the sea" or "the water," so-called because of prevalence of water-related constellation names

Virgo is second-largest constellation in night sky and biggest in zodiac

Rotation of Earth around its axis

Ecliptic is path that Sun appears to take over course of a year as Sun orbits it

Cancer, the crab, is medium-sized but very faint

Pisces, the fish, lies along celestial equator

Celestial equator lies directly above Earth's equator

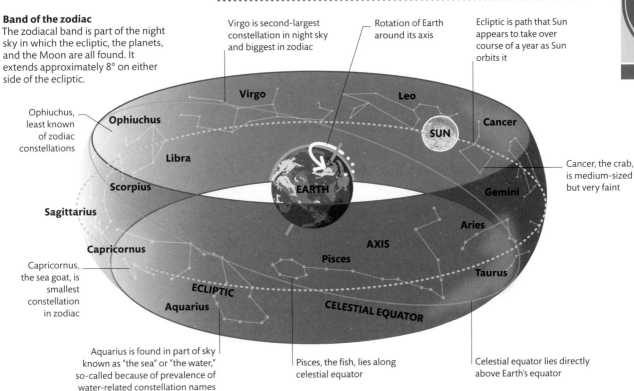

The zodiac

The 13 constellations that intersect with the path the Sun appears to trace in the sky over the course of a year are known as the constellations of the zodiac. They include the 12 "star-sign" constellations and a thirteenth, Ophiuchus, which is situated between Sagittarius and Scorpius. The zodiac comprises around one-sixth of the surface area of the celestial sphere.

THE CONSTELLATION **HYDRA** IS SO LARGE THAT IT COVERS **3 PERCENT** OF THE **ENTIRE NIGHT SKY**

BAYER DESIGNATIONS

A system of naming stars invented by German astronomer Johann Bayer in 1603 is still used today. A star is named by a Greek letter followed by the constellation name it falls within. These letters were assigned in order of brightness with the 17th-century equipment available to Bayer.

Pollux is known as Beta Geminorum by its Bayer designation, but it is now known to be the brightest star in Gemini

It is now known that Castor (Alpha Geminorum) is fainter than Pollux

GEMINI CONSTELLATION

DO THE CONSTELLATIONS CHANGE OVER TIME?

In around 50,000 years, some constellations will bear no resemblance to their current patterns. The farther a star is away from Earth, the less it will change position.

Mapping the sky

A star chart is a flattened representation of part of the celestial sphere (see p.12). A typical chart shows the names and positions of stars and constellations and often other objects, such as clusters and nebulae. Stars are usually represented by dots, with large dots for bright stars and small dots for faint stars.

How to navigate the skies

As your view of the sky depends on the hemisphere you are in and your latitude, it is important to find a chart that corresponds to your location. When looking at the night sky, the best way to orient yourself is often to find a few bright stars and constellations and then use them as pointers to other stars. A useful tool for orientation is a planisphere, which consists of a circular chart with an oval window that can be rotated to show how the sky looks at a given date and time.

Northern hemisphere
This star chart shows the constellations in the northern celestial hemisphere (dark blue circle) and up to 30° into the southern celestial hemisphere (paler blue circle). "Pointer" stars Merak and Dubhe help guide the way to Polaris, the north star.

Celestial equator is a projection of Earth's equator onto celestial sphere and zero point for declination, one of two celestial coordinates used to define star positions

Point at which celestial equator intersects ecliptic is zero point for right ascension, one of two celestial coordinates used to define positions of stars; it is measured in hours and minutes

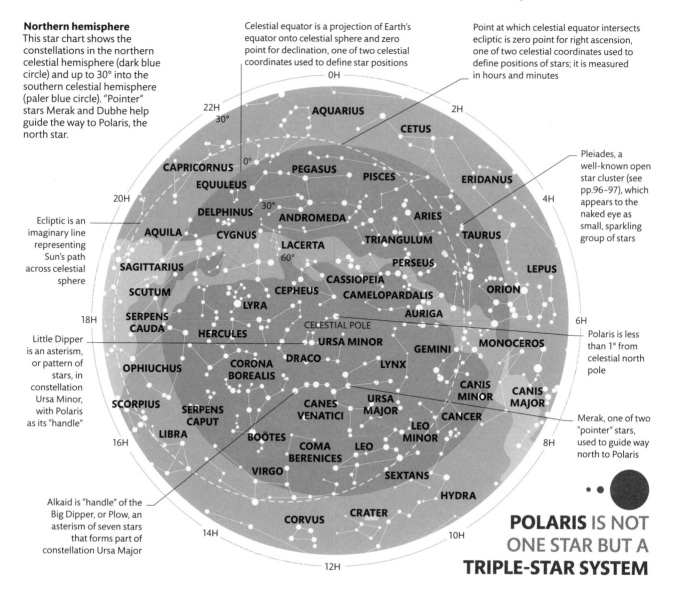

Ecliptic is an imaginary line representing Sun's path across celestial sphere

Pleiades, a well-known open star cluster (see pp.96–97), which appears to the naked eye as small, sparkling group of stars

Little Dipper is an asterism, or pattern of stars, in constellation Ursa Minor, with Polaris as its "handle"

Polaris is less than 1° from celestial north pole

Merak, one of two "pointer" stars, used to guide way north to Polaris

Alkaid is "handle" of the Big Dipper, or Plow, an asterism of seven stars that forms part of constellation Ursa Major

POLARIS IS NOT ONE STAR BUT A TRIPLE-STAR SYSTEM

HOW NEAR ARE THE CLOSEST STARS?

Proxima Centauri is the closest star to Earth, situated around 4.22 light-years away. The closest star system to us is Alpha Centauri, which lies 4.37 light-years away.

Southern hemisphere
Unlike in the northern hemisphere, in the southern hemisphere, there is no bright star at the south celestial pole, but the direction of south can be deduced using an asterism called the Southern Cross.

THE BORTLE SCALE

Light from artificial sources, especially in urban environments, obscures the view of the night sky, meaning that only the brightest objects can be seen. The greater the light pollution, the fewer stars visible in that area. The Bortle scale was created in 2001 to evaluate light pollution in given locations. It ranges from 1–9, with 1 representing the clearest skies.

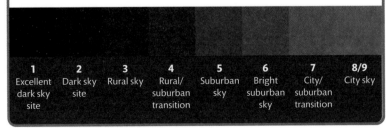

1	2	3	4	5	6	7	8/9
Excellent dark sky site	Dark sky site	Rural sky	Rural/ suburban transition	Suburban sky	Bright suburban sky	City/ suburban transition	City sky

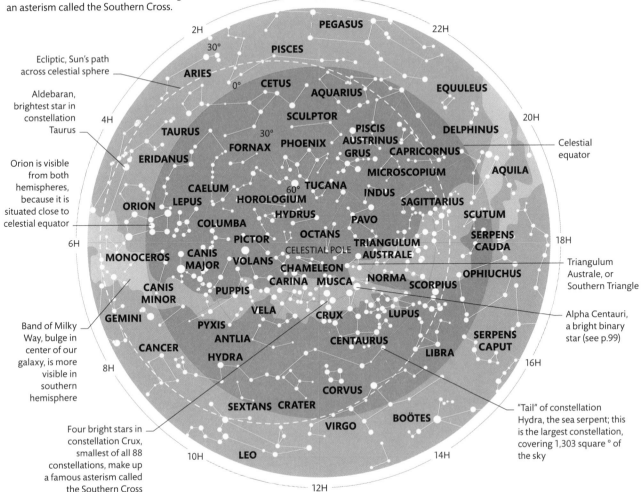

Ecliptic, Sun's path across celestial sphere

Aldebaran, brightest star in constellation Taurus

Orion is visible from both hemispheres, because it is situated close to celestial equator

Band of Milky Way, bulge in center of our galaxy, is more visible in southern hemisphere

Four bright stars in constellation Crux, smallest of all 88 constellations, make up a famous asterism called the Southern Cross

Celestial equator

Triangulum Australe, or Southern Triangle

Alpha Centauri, a bright binary star (see p.99)

"Tail" of constellation Hydra, the sea serpent; this is the largest constellation, covering 1,303 square ° of the sky

Telescopes

It is possible to view many objects in the night sky with the naked eye. However, to study these in more detail and to view fainter objects requires a piece of equipment capable of collecting and focusing light to produce a magnified image. Telescopes do this in two ways: by using either mirrors or lenses.

THE **REFLECTING TELESCOPE** WAS INVENTED BY SIR **ISAAC NEWTON** IN **1668**

Reflecting telescopes

Telescopes work by gathering as much light as they can and then focusing it to one point. This results in a bright, clear picture of a distant object. Reflecting telescopes focus the light from an object using flat or curved mirrors. One benefit of reflecting telescopes over refracting ones is that the mirrors can be made very large without becoming too heavy, unlike lenses.

How a reflecting telescope works
A telescope's magnification depends on the focal length—the distance from a lens or mirror to the point where the light rays meet (the focal point). The longer the focal length, the greater the magnification.

4 Eyepiece
The eyepiece lens magnifies the image. The shorter the focal length of the lens, the larger the image appears.

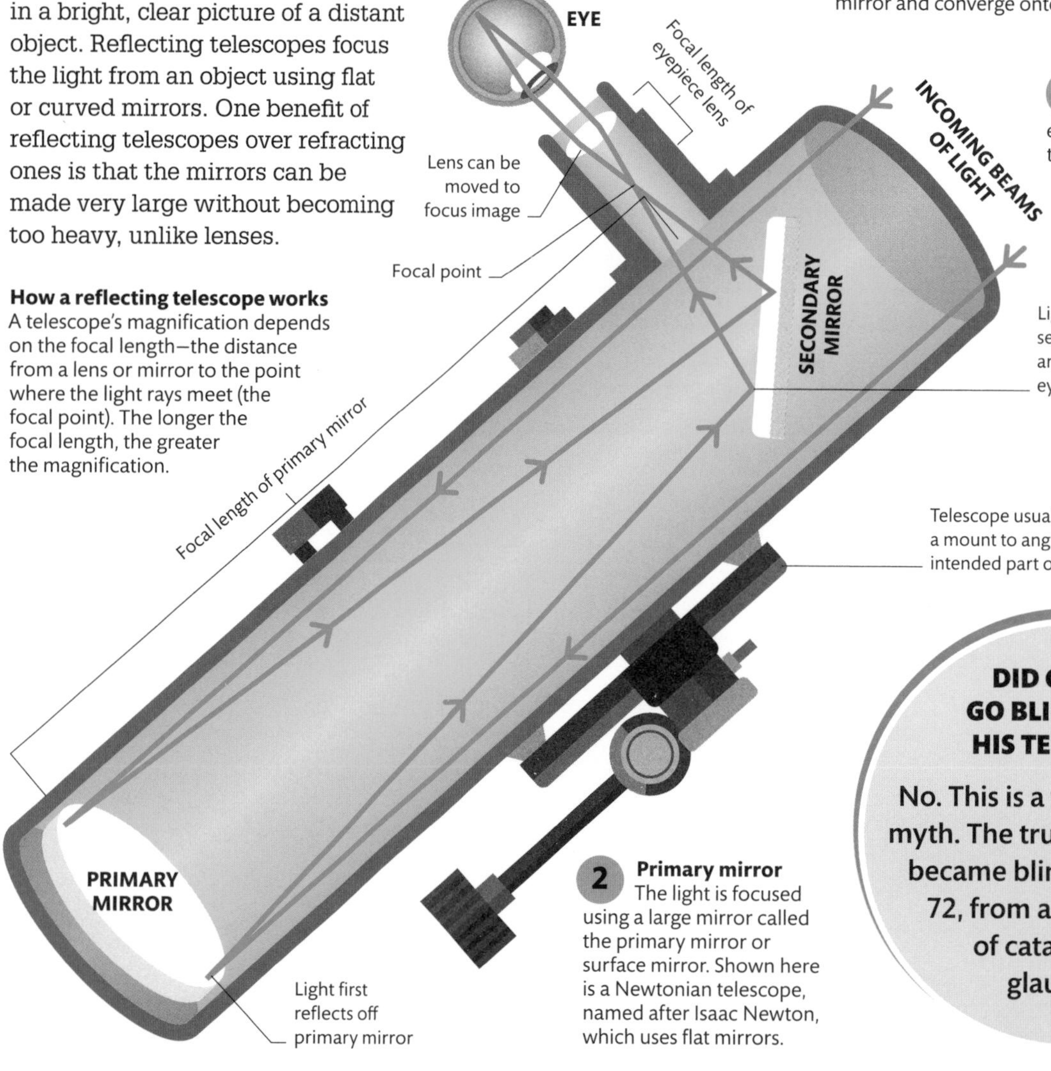

EYE

Focal length of eyepiece lens

Lens can be moved to focus image

Focal point

Focal length of primary mirror

3 Secondary mirror
The beams of light that reflect off the primary mirror are directed toward the smaller, secondary mirror. From there, the light beams reflect off different parts of the mirror and converge onto a focal point.

INCOMING BEAMS OF LIGHT

1 Incoming light
Parallel light rays enter through the top of the telescope.

SECONDARY MIRROR

Light reflects off secondary mirror and travels into eyepiece

Telescope usually sits on a mount to angle it at intended part of night sky

PRIMARY MIRROR

Light first reflects off primary mirror

2 Primary mirror
The light is focused using a large mirror called the primary mirror or surface mirror. Shown here is a Newtonian telescope, named after Isaac Newton, which uses flat mirrors.

DID GALILEO GO BLIND USING HIS TELESCOPE?

No. This is a widely believed myth. The truth is that Galileo became blind at the age of 72, from a combination of cataracts and glaucoma.

Refracting telescopes

Refracting telescopes use lenses to produce a magnified image. While these telescopes are more robust and need less maintenance than reflecting telescopes, the lenses have to be very large to see distant objects, which makes them heavy. This also means that any slight imperfection in the lens will have a big impact on the final image. They also suffer from defects like chromatic aberration, where colors are bent to different extents by the lens due to their different wavelengths.

How a refracting telescope works
A simple refracting telescope can be created using two lenses, both convex. The biggest lens is an objective lens, which focuses light from a distant object.

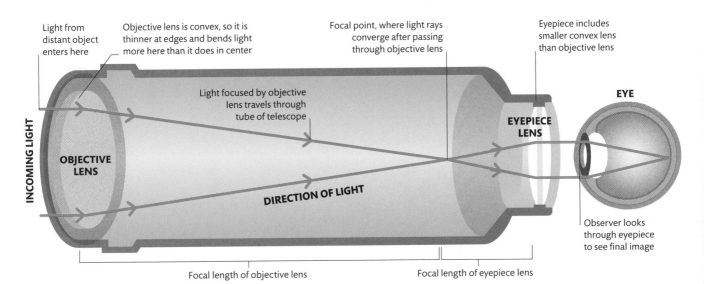

Light from distant object enters here

Objective lens is convex, so it is thinner at edges and bends light more here than it does in center

Focal point, where light rays converge after passing through objective lens

Eyepiece includes smaller convex lens than objective lens

Light focused by objective lens travels through tube of telescope

EYE

INCOMING LIGHT

OBJECTIVE LENS

EYEPIECE LENS

DIRECTION OF LIGHT

Observer looks through eyepiece to see final image

Focal length of objective lens

Focal length of eyepiece lens

1 Objective lens
Parallel light rays enter the telescope and hit the objective lens. The lens is convex, which means it focuses the light to a point. The bigger the objective lens, the more the telescope is able to magnify the object.

2 Focal point
This is the point where the focused rays of light come together after passing through the objective lens. Here, an image is at its sharpest. After this point, the light disperses again.

3 Eyepiece
A small lens is used to refract the light that has passed through the objective lens. Light rays that pass through the lens exit in parallel, creating a virtual image in the eyepiece.

TELESCOPE MOUNTS

Telescopes are usually placed on mounts to keep them steady and help the viewer find objects in the sky. There are two main types of telescope mount: alt-azimuth and equatorial. An alt-azimuth mount uses two axes of rotation, both of which need to be moved to track a celestial object. An equatorial mount also uses two axes but has one axis aligned so it points to the celestial pole (see pp.12–13).

Telescope tilts up and down

Telescope moves from side to side

ALT-AZIMUTH MOUNT

Telescope tilts up and down

Axis already tilted toward celestial pole, so observer only needs to move telescope up and down

EQUATORIAL MOUNT

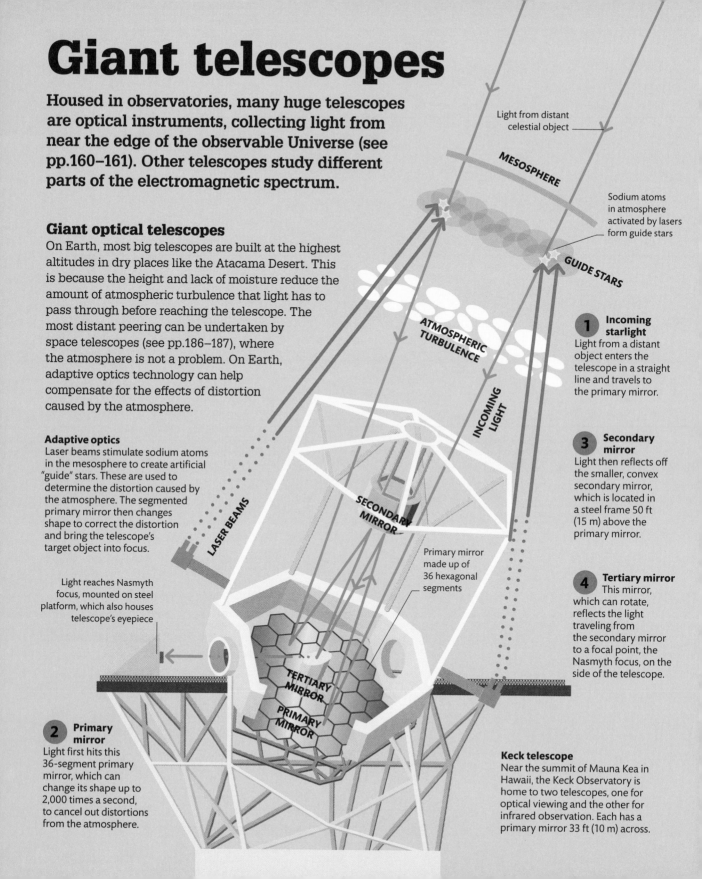

Giant telescopes

Housed in observatories, many huge telescopes are optical instruments, collecting light from near the edge of the observable Universe (see pp.160–161). Other telescopes study different parts of the electromagnetic spectrum.

Giant optical telescopes

On Earth, most big telescopes are built at the highest altitudes in dry places like the Atacama Desert. This is because the height and lack of moisture reduce the amount of atmospheric turbulence that light has to pass through before reaching the telescope. The most distant peering can be undertaken by space telescopes (see pp.186–187), where the atmosphere is not a problem. On Earth, adaptive optics technology can help compensate for the effects of distortion caused by the atmosphere.

Adaptive optics

Laser beams stimulate sodium atoms in the mesosphere to create artificial "guide" stars. These are used to determine the distortion caused by the atmosphere. The segmented primary mirror then changes shape to correct the distortion and bring the telescope's target object into focus.

Light reaches Nasmyth focus, mounted on steel platform, which also houses telescope's eyepiece

LASER BEAMS

2 Primary mirror
Light first hits this 36-segment primary mirror, which can change its shape up to 2,000 times a second, to cancel out distortions from the atmosphere.

Light from distant celestial object

MESOSPHERE

Sodium atoms in atmosphere activated by lasers form guide stars

GUIDE STARS

ATMOSPHERIC TURBULENCE

INCOMING LIGHT

SECONDARY MIRROR

Primary mirror made up of 36 hexagonal segments

TERTIARY MIRROR

PRIMARY MIRROR

1 Incoming starlight
Light from a distant object enters the telescope in a straight line and travels to the primary mirror.

3 Secondary mirror
Light then reflects off the smaller, convex secondary mirror, which is located in a steel frame 50 ft (15 m) above the primary mirror.

4 Tertiary mirror
This mirror, which can rotate, reflects the light traveling from the secondary mirror to a focal point, the Nasmyth focus, on the side of the telescope.

Keck telescope

Near the summit of Mauna Kea in Hawaii, the Keck Observatory is home to two telescopes, one for optical viewing and the other for infrared observation. Each has a primary mirror 33 ft (10 m) across.

Signal bounces off concave angle of dish

1 Incoming signal
Incoming radio waves reflect off the main dish, which is usually large to gather as much signal as possible. This is partly because radio signals from distant sources are often very weak.

SUBREFLECTOR

Incoming radio signal from celestial object

FEED HORN

MAIN DISH

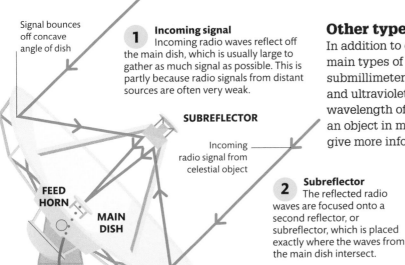

Other types of telescope

In addition to optical telescopes, there are four main types of telescope: radio telescopes, submillimeter telescopes, infrared telescopes, and ultraviolet telescopes. Each is named for the wavelength of radiation that it detects. Looking at an object in multiple parts of the spectrum can give more information than just a single region.

2 Subreflector
The reflected radio waves are focused onto a second reflector, or subreflector, which is placed exactly where the waves from the main dish intersect.

How a radio telescope works
Radio telescopes are specifically designed to receive long-wavelength radio waves from space. They typically feature a large parabolic dish that reflects the radio waves to a subreflector and on to a receiver.

Receiver transmits signal to computer

Computer and recording equipment interprets signal

Signal travels via fiber-optic cable

RECEIVER

3 Feed horn
After bouncing off the subreflector, the signal travels through the feed horn in the center of the dish to the receiver.

4 Receiver
The receiver features an amplifier, which increases the signal strength. Then the signal travels to a computer.

5 Computer
Signals are stored on a computer and are either processed there or sent on for analysis using sophisticated software.

WHAT IS THE WORLD'S HIGHEST OBSERVATORY?

The University of Tokyo's Atacama Observatory, on the summit of Cerro Chajnantor in Chile, is located at a height of 18,500 ft (5,640 m).

THE **KECK TELESCOPE** CAPTURED THE **FIRST IMAGE** OF AN **EXTRASOLAR PLANETARY SYSTEM** IN 2008

ASTRONOMICAL INTERFEROMETRY

An astronomical interferometer combines the light or radio signals from two or more telescopes. This allows astronomers to examine a celestial object in more detail, as though it is being observed using mirrors or antennas measuring hundreds of feet (m) in diameter. It is achieved by setting up arrays of telescopes that observe an object at the same time. A digital correlator processes the signals and allows for the time lag between the telescopes.

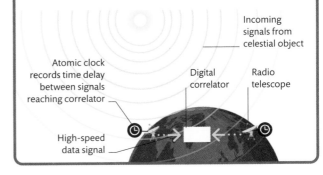

Incoming signals from celestial object

Atomic clock records time delay between signals reaching correlator

Digital correlator

Radio telescope

High-speed data signal

Spectroscopy

Astronomers can identify what elements or molecules are present in a star or other celestial object by studying the light that it emits or absorbs. This is undertaken using a technique called spectroscopy, which splits electromagnetic radiation into separate wavelengths.

What stars are made of

Visible light is one part of a spectrum of electromagnetic radiation (see pp.152–153). Elements emit different wavelengths of light, depending on their inherent energy levels. Because we know the wavelengths that correspond to particular elements, we can use instruments to analyze light to find out what stars and other celestial objects, including nebulae (see pp.94–95) and black holes, are made of. One such instrument is a spectroscope, which focuses a beam of light at a prism to separate it into its constituent wavelengths.

STAR

Longer wavelengths of light, like red and orange light, bend the least and carry the least energy

Spectroscope prism splits visible starlight into its many wavelengths

LIGHT FROM STAR

Light wave traveling from star into spectroscope prism

SPECTROSCOPE PRISM

Blue and violet light have shorter wavelengths, so they bend more and produce higher amounts of energy

How a spectroscope works
Starlight travels into a prism, a transparent optical device that bends light. As light enters the prism, it slows down, but each wavelength, which corresponds to a different color, slows down to a different extent. The wavelengths exit the prism at different points, producing a rainbow of colors.

SPECTROGRAPHS

Spectrographs are more sophisticated instruments than spectroscopes. They use thin slits, mirrors, and a diffraction grating—an opaque screen scored with many transparent parallel lines—to separate the light at a more detailed level. Instead of a rainbow, the output is a spectrum in which the light is separated into individual wavelengths. Increasingly, astronomers use a technique called multiobject spectroscopy, in which they study the spectra from more than one celestial object within the field of view of the instrument at the same time.

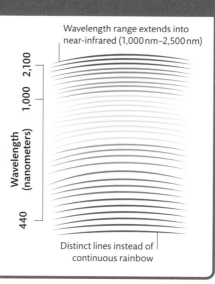

Wavelength range extends into near-infrared (1,000 nm–2,500 nm)

Wavelength (nanometers)

2,100

1,000

440

Distinct lines instead of continuous rainbow

SPECTROGRAPHS CAN REVEAL HOW QUICKLY STARS MOVE

WHO FIRST ANALYZED STARLIGHT?

Physicist Joseph von Fraunhofer invented the spectroscope in 1814 and used it to study the Sun's spectra. The absorption lines he found are named in his honor.

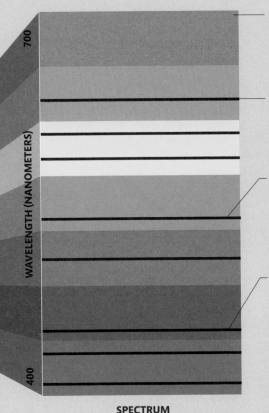

WAVELENGTH (NANOMETERS)

700

400

In electromagnetic spectrum, visible light ranges from red to violet; our eyes can detect wavelengths from around 400 to 700 nanometers

Each element produces its own unique pattern of black absorption lines, enabling astronomers to detect their presence in a star

Width of lines appearing on a spectrum varies depending on instrument used and temperature of material

In this absorption spectrum (see below), black lines are dips where specific wavelengths of light are missing

Unique chemical fingerprint
Each star has its own spectrum, with each spectrum revealing exactly what materials are present in the star and its atmosphere. Spectra can help astronomers tell stars apart and reveal what stars have in common.

SPECTRUM

Starlight from Eta Carinae behind | Dumbbell-shaped nebula

Stars with unusual spectra
Analyzing the spectrum of the double supergiant Eta Carinae, hidden from direct view by a nebula formed from ejected stellar material 170 years ago, shows that the nebula is rich in nickel and iron.

HELIUM WAS ONLY **DISCOVERED** IN **1868** WHILE ASTRONOMERS WERE STUDYING THE **SPECTRA OF THE SUN**

Types of spectrum

Depending on the object viewed, a spectroscope can produce three different types of spectrum. A continuous spectrum is created by a solid or a hot, dense gas, and it looks like a rainbow, with all the wavelengths of visible light represented. An absorption spectrum can be produced by a hot object like a star seen through a cooler gas. This type of spectrum is caused by atoms in a gas cloud absorbing the star's energy at specific wavelengths and then reemitting them randomly. An emission spectrum is produced by a hot, low-density gas, which emits light at specific wavelengths only. It appears as a series of bright lines, each corresponding to a wavelength at which emission takes place.

Distinctive patterns
The three types of spectrum produce identifiable patterns. An absorption spectrum looks like a continuous spectrum minus the emission lines. Light from the Sun is very nearly a continuous spectrum, but gases in its atmosphere absorb certain wavelengths of light, producing an absorption spectrum.

Spectrum appears as an unbroken rainbow

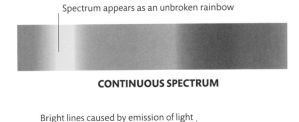

CONTINUOUS SPECTRUM

Bright lines caused by emission of light

EMISSION SPECTRUM

Dark lines caused by absorption of light

ABSORPTION SPECTRUM

Rocks from space

Many bodies of rock—as well as ice and metal—orbit the Sun. Some, such as comets and asteroids, are vast. Meteoroids are much smaller, and when they enter Earth's atmosphere, they are known as meteors, or shooting stars. The few meteors that are not completely vaporized and go on to strike Earth's surface are called meteorites.

COMET

Small nucleus of ice and dust surrounded by bright cloud, or coma, of gas and dust

ASTEROID

Solid body made up of rocky materials and metals; formed from remnants of failed planet formation

Traveling into the atmosphere
After hurtling through the vacuum of space, objects suddenly slow down very rapidly upon entering Earth's atmosphere. The friction caused by the various layers of Earth's atmosphere burns the solid material, usually vaporizing it.

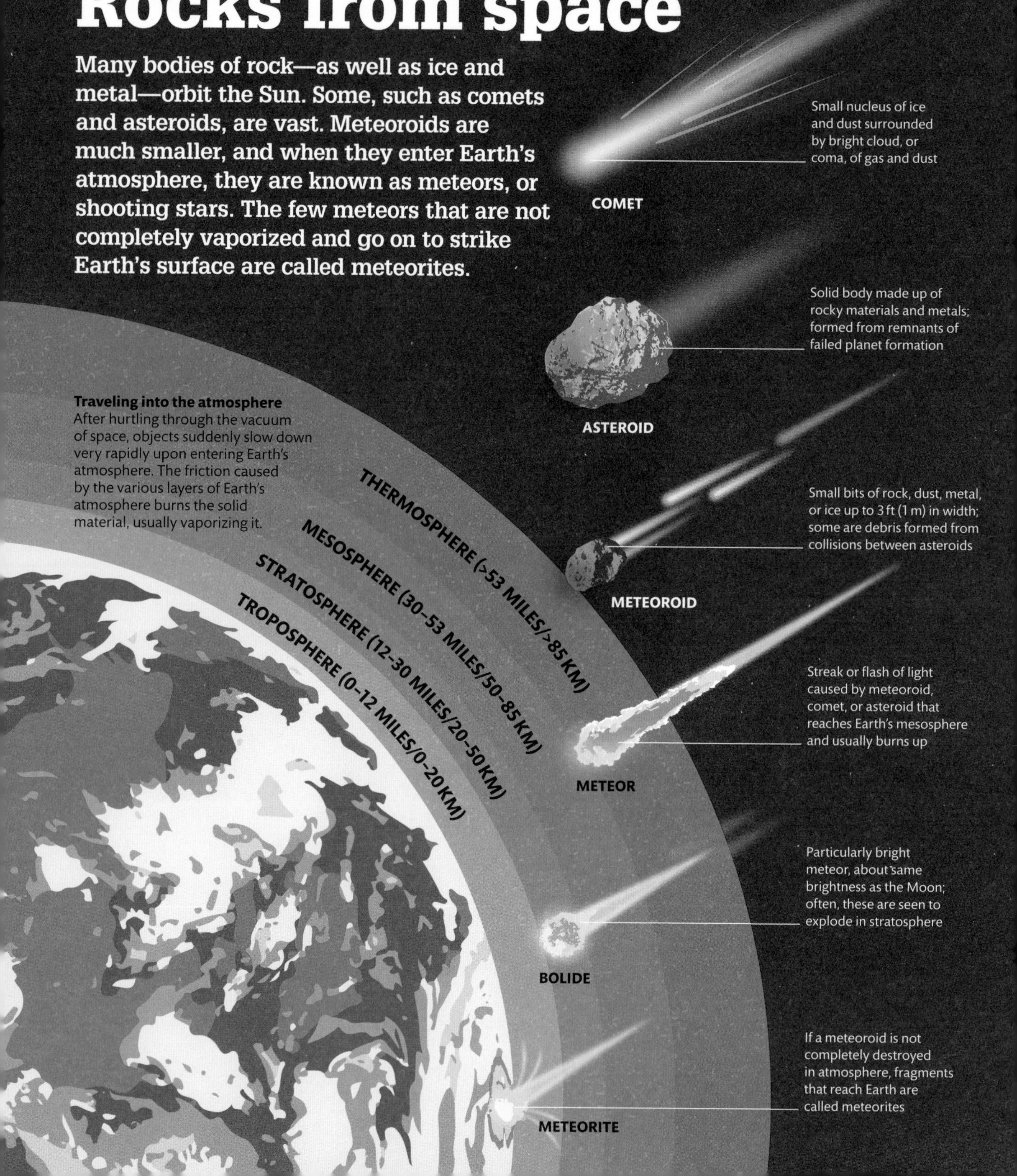

THERMOSPHERE (>53 MILES/>85 KM)

MESOSPHERE (30–53 MILES/50–85 KM)

STRATOSPHERE (12–30 MILES/20–50 KM)

TROPOSPHERE (0–12 MILES/0–20 KM)

METEOROID

Small bits of rock, dust, metal, or ice up to 3 ft (1 m) in width; some are debris formed from collisions between asteroids

METEOR

Streak or flash of light caused by meteoroid, comet, or asteroid that reaches Earth's mesosphere and usually burns up

BOLIDE

Particularly bright meteor, about same brightness as the Moon; often, these are seen to explode in stratosphere

METEORITE

If a meteoroid is not completely destroyed in atmosphere, fragments that reach Earth are called meteorites

Types of rock

There are lots of fragments of rock moving around the Solar System, left over from when the planets and moons were forming. Objects made of rock, up to 3 ft (1 m) in size, are termed meteoroids. Rocky objects larger than this but too small to be spherical like a planet are generally asteroids or comets. Asteroids can be up to 600 miles (1,000 km) in size, while comets are smaller, up to around 25 miles (40 km). Most asteroids are in the Main Belt between Mars and Jupiter (see pp.60–61). Comets originate much farther from Earth, which makes them cold enough to contain ice. When parts of these objects enter Earth's atmosphere and burn up, they create meteors.

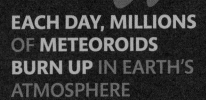

EACH DAY, MILLIONS OF METEOROIDS BURN UP IN EARTH'S ATMOSPHERE

WHAT'S THE BIGGEST RECORDED METEORITE TO HIT EARTH?

The biggest intact meteorite is the Hoba meteorite, found in Namibia. It is thought to have fallen to Earth 80,000 years ago and weighs 60 tons (12,000 lb).

Meteorites

Meteorites are divided into three main types: iron, stony, and stony-iron. Meteorites often feature a burned, shiny exterior created as their outer surface melts when passing through the atmosphere. Some meteorites comprise material that originally formed the rocky planets, thus providing a glimpse into the conditions at the beginning of our Solar System.

TYPES OF METEORITE

Meteorite type	Composition	Origin	Percentage of meteorites
IRON	Composed mainly of iron-nickel alloy and small amounts of other minerals.	Thought to be the cores of asteroids that melted early in their history.	5.4 percent
STONY	Silicate minerals; they are divided into two groups: achondrites and chondrites. Chondrites contain once-molten grains called chondrules.	Achondrites formed by melting of parent asteroids; chondrites formed in the primitive Solar System from dust, ice, and grit.	93.3 percent
STONY-IRON	Roughly equal amounts of metal and silicate crystals; they are divided into two groups: pallasites and mesosiderites.	Pallasites formed between a metal core and an outer silicate mantle; mesosiderites form through a collision between asteroids.	1.3 percent

METEOR SHOWERS

Comets are always losing bits of themselves, leaving behind a trail in their wake. When Earth's orbit of the Sun brings us through that trail, we experience a meteor shower. During these periods, it can be possible to witness tens to hundreds of meteors radiating from a common point in the night sky in just an hour. Meteor showers are usually named after a star or constellation near where the meteors originate from in the sky.

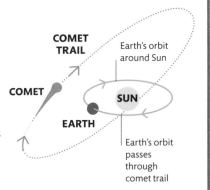

COMET TRAIL

Earth's orbit around Sun

COMET

SUN

EARTH

Earth's orbit passes through comet trail

Particles from space

Space is almost but not quite a vacuum. There are many different types of particles traveling through space, including a stream of charged particles emanating from the Sun. The majority of these particles that approach Earth are deflected by our planet's magnetic field. However, some particles can get through and interact with our atmosphere.

CHARGED PARTICLES THAT CAUSE **AURORAE TRAVEL** AT ABOUT **250 MILES (400 KM) PER SECOND**

SOLAR WIND

SUN

Sunspots are dark, relatively cool regions on Sun's surface caused by concentrations of Sun's magnetic field

Composition of the solar wind
The solar wind is a mixture of particles released from the Sun's upper atmosphere, or corona. It is made up mostly of charged particles, or ions, of hydrogen; helium nuclei; and heavier ions, including carbon, nitrogen, and oxygen.

Prominences are loops of hydrogen and helium in a plasma state that reach out into space, although they remain attached to photosphere (Sun's visible surface)

Solar wind takes between two and four days to reach Earth

SOLAR WIND

Corona (Sun's exterior layer) extends into space

Cosmic rays

Although given the name cosmic rays, these are not really rays at all. They are high-energy subatomic particles that travel from the Solar System or beyond. Most of these, 89 percent, are positively charged particles called protons, or hydrogen nuclei. (There is one proton in a hydrogen nucleus.) Another 10 percent are helium nuclei, comprising two protons and two neutrons, and the remainder are nuclei of heavier elements. These particles travel through space at close to the speed of light. Just how they reach high enough energies to move this fast is an unsolved puzzle.

WHY IS AN AURORA COLORFUL?

The color is caused by the type of atoms in Earth's atmosphere and the height at which solar wind particles hit them. Green lights are caused by oxygen particles 60 miles (100 km) up.

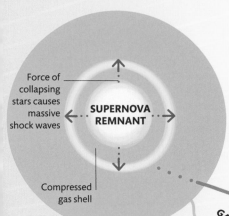

Force of collapsing stars causes massive shock waves

SUPERNOVA REMNANT

Compressed gas shell

The solar wind

Charged particles from the Sun, known as the solar wind, make up the lowest-energy cosmic rays that reach Earth. Aurorae are caused by these particles entering Earth's atmosphere and colliding with gas particles in the air. This provides the gas particles with extra energy and excites electrons within them to a higher-energy state. This state is unstable, so the electrons will return to their previous state, releasing the energy as a photon, or a particle of light.

COSMIC RAY

GAMMA RAY

Deflected charged particles pass through magnetosphere at areas of weaker field strength called cusps; from there, they travel to Earth's magnetically charged poles

Supernova sources

When huge stars explode, they create shock waves, which are thought to accelerate charged particles and gamma rays (see pp.152–153) to very high energies. While charged particles are deflected by Earth's magnetic field away from Earth, electrically neutral gamma rays are not.

Spherical outer radiation belt traps incoming solar wind particles

Defending Earth

Electricity in Earth's molten iron core generates a magnetic field, which forms a protective bubble around the planet. This helps protect us from charged particles, many of which are deflected around Earth.

Aurorae manifest in huge rings, called auroral ovals, above Earth's magnetic poles

Aurorae around south pole are known as aurora australis, or southern lights

Inner radiation belt, consisting mainly of highly energetic protons

Most particles are deflected away from Earth by magnetic field

Magnetopause, the edge of Earth's magnetic field

SPACE WEATHER

The magnetic activity on the surface of the Sun creates a type of weather called space weather. Mass ejections from the Sun's corona, for example, can create geomagnetic storms. In the most extreme cases, these can impact orbiting satellites and even power grids on Earth's surface.

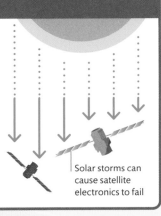

Solar storms can cause satellite electronics to fail

PARTICLE AIR SHOWER

PROTON

ATMOSPHERIC MOLECULES

PION **PION** **PION**

NEUTRON
ANTINEUTRINO

MUON **MUON**

PHOTON **PHOTON**

ELECTRON **POSITRON** **ELECTRON** **POSITRON**

Descent through Earth's atmosphere

Cosmic rays interact with molecules in Earth's atmosphere, producing subatomic particles called pions. In turn, these may decay or collide with other particles in the air and create a cascade of further particles.

Looking for aliens

The question of whether life exists beyond Earth has captured the imaginations of humans for centuries. Attempts to identify extraterrestrial life principally involve launching probes into space and scanning for radio signals that may have been sent by aliens.

WHAT IS A FAST RADIO BURST?

Fast radio bursts are mysterious pulses of powerful radio waves that last for only a few milliseconds and usually come from distant galaxies. Their origin is unknown.

Trying to make contact

In 1974, radio signals were transmitted for the first time to try to make contact with extraterrestrial life. The Search for Extraterrestrial Intelligence (SETI) Institute, launched in 1985, built on these efforts. Later developments include the completion of the Five-hundred-meter Spherical Aperture Telescope (FAST) in 2019. One of its functions is to listen for alien radio signals.

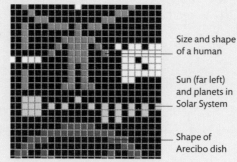

Size and shape of a human

Sun (far left) and planets in Solar System

Shape of Arecibo dish

Arecibo message
In 1974, a radio message was sent from the Arecibo Observatory to star cluster M13. It included data about humanity and Earth.

THE **FAST TELESCOPE** HAS A **COLLECTING AREA** EQUIVALENT TO **750** TENNIS COURTS

Adjustable panels
The reflector itself is too large to move, but its 4,500 triangular panels can be adjusted, forming a kind of flexible mirror that can be deformed to widen the search area.

Each panel weighs around 1,000 lb (450 kg)

Panels made from perforated aluminum

SCALE

INCOMING RADIO WAVE

Network of steel cables support receiver cabin

Receiver cabin, containing multiple-beam and band radio receivers

FAST telescope
FAST is the largest radio telescope in the world. It is located in a natural basin in a mountainous region of China, which protects it from radio-wave interference. It could be used to scan for radio signals coming from distant exoplanets, potential locations for alien life.

MAIN REFLECTOR

Cosmic quiet zone

The "water hole" is a band of the electromagnetic spectrum between 1,420 and 1,640 MHz in which interference is minimal. The frequency range is associated with emissions from hydrogen atoms and hydroxyl particles, which together constitute water. It is a popular listening frequency for radio telescopes.

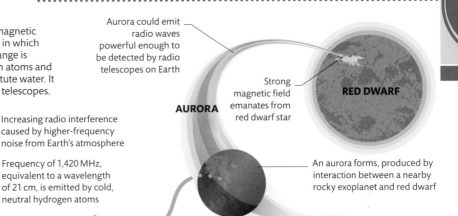

Aurora could emit radio waves powerful enough to be detected by radio telescopes on Earth

RED DWARF

AURORA

Strong magnetic field emanates from red dwarf star

Increasing radio interference caused by higher-frequency noise from Earth's atmosphere

Frequency of 1,420 MHz, equivalent to a wavelength of 21 cm, is emitted by cold, neutral hydrogen atoms

An aurora forms, produced by interaction between a nearby rocky exoplanet and red dwarf

EXOPLANET

How SETI@Home worked

This citizen science experiment, which ran from 1999 to 2020, allowed anyone with a computer and an internet connection to help in the search for extraterrestrial life. Users installed a free program, which then downloaded and analyzed 107-second units of data collected from radio telescopes.

Listening for aliens

One way of trying to find aliens is by listening for signals from intelligent life, sent out on purpose to make contact with other intelligent life forms. This is done by searching for electromagnetic radiation in the radio frequency and ruling out any other possible sources for that radiation. SETI@Home was a unique program that has been at the forefront of this endeavor. The collected data is still being analyzed.

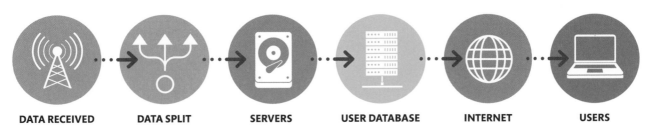

DATA RECEIVED **DATA SPLIT** **SERVERS** **USER DATABASE** **INTERNET** **USERS**

THE DRAKE EQUATION

This is an equation used to estimate not only how likely it is that life exists outside our planet, but also the odds of humans being able to find intelligent life in the Universe. First proposed by radio astronomer Frank Drake in 1961, the equation calculates the number of civilizations potentially capable of communication by multiplying several variables.

Number of advanced civilizations in Milky Way

Fraction of stars with planetary systems

Fraction of those worlds that give rise to life

Fraction of civilizations with communications technology

$$N = R_* \times f_p \times n_e \times f_e \times f_i \times f_c \times L$$

Rate of formation of stars in galaxy

Number of life-supporting worlds per planetary system

Fraction of those worlds with intelligent life

Average lifetime of communicating civilization

THE SOLAR SYSTEM

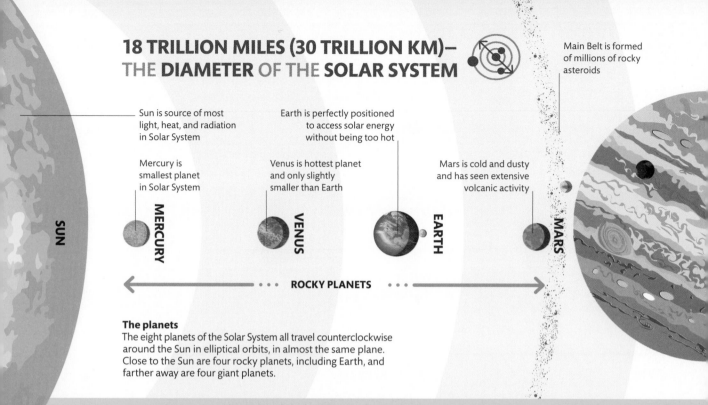

18 TRILLION MILES (30 TRILLION KM)— THE DIAMETER OF THE SOLAR SYSTEM

Main Belt is formed of millions of rocky asteroids

Sun is source of most light, heat, and radiation in Solar System

Earth is perfectly positioned to access solar energy without being too hot

Mercury is smallest planet in Solar System

Venus is hottest planet and only slightly smaller than Earth

Mars is cold and dusty and has seen extensive volcanic activity

SUN

MERCURY

VENUS

EARTH

MARS

ROCKY PLANETS

The planets
The eight planets of the Solar System all travel counterclockwise around the Sun in elliptical orbits, in almost the same plane. Close to the Sun are four rocky planets, including Earth, and farther away are four giant planets.

Structure of the Solar System

The Solar System is structured around the Sun, with a clear distinction between small, rocky bodies close to the Sun and giant gas and ice planets much farther away.

Objects in the Solar System

The Solar System comprises all of the objects held by the Sun's powerful gravitational pull. The largest such objects are the eight known planets, which have over 200 moons between them. Rocky asteroids and icy comets race through the spaces between the planets and the five confirmed dwarf planets. The Solar System extends to the edge of the Oort Cloud (see pp.84–85)—around 100,000 times the distance between Earth and the Sun. It is just one of hundreds of billions of similar structures embedded in the vast stellar metropolis known as the Milky Way Galaxy.

THE ICE LINE

The ice line marks the point in a forming planetary system where temperatures drop below the freezing point of water, ammonia, and methane. Beyond this line, icy material gathers to form giant planets. Closer to the star, only rock and metal can withstand the heat.

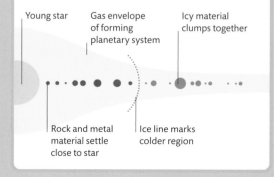

Young star

Gas envelope of forming planetary system

Icy material clumps together

Rock and metal material settle close to star

Ice line marks colder region

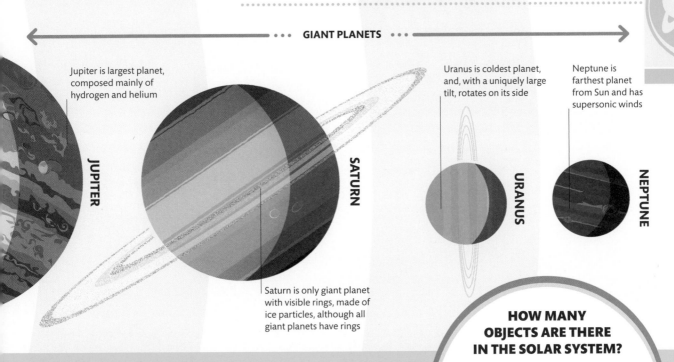

GIANT PLANETS

Jupiter is largest planet, composed mainly of hydrogen and helium

JUPITER

SATURN

Uranus is coldest planet, and, with a uniquely large tilt, rotates on its side

URANUS

Neptune is farthest planet from Sun and has supersonic winds

NEPTUNE

Saturn is only giant planet with visible rings, made of ice particles, although all giant planets have rings

Kepler's laws of planetary motion

German astronomer Johannes Kepler used detailed observations of the movements of the planets to formulate three mathematical laws. Later, Isaac Newton showed how Kepler's laws followed naturally from his law of universal gravitation. The three laws describe the shapes of orbits and how the speed of motion is affected by the distance from the Sun.

HOW MANY OBJECTS ARE THERE IN THE SOLAR SYSTEM?

The exact number is unknown, but more than half a million Solar System bodies have official names, and there are at least another 300,000 yet to be named.

Planet orbits Sun in ellipse

Planet is always at same combined distance from two focus points

Sun is at focus point

Secondary focus point

Planet travels faster close to Sun

Shaded portions have equal areas

Sun

100-day period

100-day period

Mars makes partial orbit in one Earth year

Jupiter completes only a fraction of orbit after one Earth year

Sun

Earth orbits Sun in one year

Saturn's full orbit takes 29 Earth years

Law 1
Kepler's first law states that the orbit of a planet is an ellipse, with two focus points and the Sun at one of these focus points. The more elliptical an orbit is, the more orbital eccentricity it is said to have.

Law 2
Kepler noticed that a planet speeds up when it is close to the Sun and slows down when it is farther away. He found that the line from the Sun to the planet sweeps out equal areas in equal periods of time.

Law 3
Planets take longer to orbit the farther away they are from the Sun. Kepler found a simple formula that links the orbital periods of the planets with the size of their orbits.

Birth of the Solar System

The Solar System formed about 4.5 billion years ago. By studying young star systems in the Milky Way and running computer simulations, astronomers have begun to understand how the Solar System probably came into being.

The solar nebula

The most widely accepted idea of how the Solar System formed starts with the birth of the Sun, when a ball of gas and dust, called a core, was pulled together by gravity inside a giant molecular cloud, possibly triggered by a nearby exploding star (see pp.92–93). As the core collapsed, more material was drawn in, adding to its central density and causing it to spin increasingly fast. A flat protoplanetary disk of gas and dust, called a solar nebula, grew around the newly formed Sun at the center. Over millions of years, gravity continued to draw the disk material together, creating the system of asteroids, moons, and planets that now orbit the Sun.

WHICH PLANET FORMED FIRST?

Astronomers believe that the gas giant Jupiter was the first planet to form and that it then influenced the way other planets formed. The rocky planets may have formed last.

0.01 PERCENT
OF THE **NEBULA MATERIAL** ENDED UP IN THE **PLANETS**

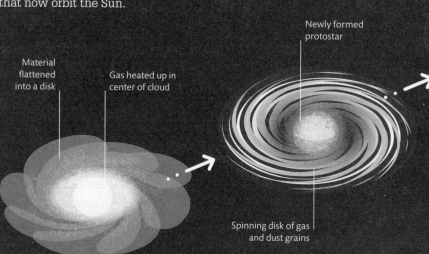

Young Sun shone brightly

Material flattened into a disk

Gas heated up in center of cloud

Newly formed protostar

Disk material clumped into planetesimals

Spinning disk of gas and dust grains

1 Core contracted
A rotating clump of material, pulled together by gravity, contracted inside an interstellar cloud. The center became hotter and denser, and a disk formed around it.

2 Protostar generated energy
Nuclear fusion started, forming a protostar. Its energy counteracted gravity, stopping the protostar from collapsing further. Grains of dust formed in the spinning disk.

3 Planetesimals formed
Disk material clumped into small bodies called planetesimals. Some material close to the star evaporated, leaving heavy elements such as iron and nickel. A solar wind blew gas farther away.

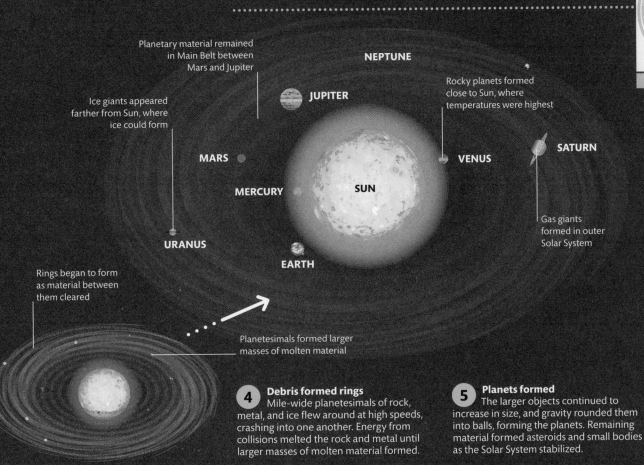

Planetary material remained
in Main Belt between
Mars and Jupiter

NEPTUNE

JUPITER

Rocky planets formed
close to Sun, where
temperatures were highest

Ice giants appeared
farther from Sun, where
ice could form

SATURN

MARS

VENUS

MERCURY

SUN

Gas giants
formed in outer
Solar System

URANUS

EARTH

Rings began to form
as material between
them cleared

Planetesimals formed larger
masses of molten material

4 **Debris formed rings**
Mile-wide planetesimals of rock,
metal, and ice flew around at high speeds,
crashing into one another. Energy from
collisions melted the rock and metal until
larger masses of molten material formed.

5 **Planets formed**
The larger objects continued to
increase in size, and gravity rounded them
into balls, forming the planets. Remaining
material formed asteroids and small bodies
as the Solar System stabilized.

Planetary migration

It took millions of years for the
Solar System to settle into its
present configuration. The
newly formed planets migrated
as they interacted with each
other and debris remaining from
their formation. This process
also depleted the Main Belt and
the Kuiper Belt beyond Neptune
(see pp.82–83) by spreading
debris far and wide.

Altered orbits
Models of planetary migration suggest that
Jupiter moved inward, while Saturn, Uranus,
and Neptune—energized by the scattering of
smaller bodies—edged farther out. Neptune
and Uranus even swapped position.

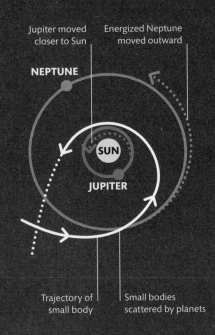

Jupiter moved
closer to Sun

Energized Neptune
moved outward

NEPTUNE

SUN

JUPITER

Trajectory of
small body

Small bodies
scattered by planets

PROTOPLANETARY DISKS

New solar systems form in flat,
dusty disks called protoplanetary
disks that swirl around newly
formed stars. Clumps of dust
appear where planets are forming.

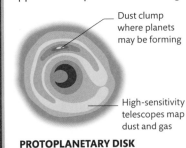

Dust clump
where planets
may be forming

High-sensitivity
telescopes map
dust and gas

PROTOPLANETARY DISK

The Sun

The Sun is an enormous nuclear powerhouse at the heart of our Solar System. It provides the gravitational force that binds the Solar System together, and its energy floods the planets with heat and light.

Inside the Sun

Solar energy begins its odyssey deep in the core of the Sun. The crush of gravity sends temperatures soaring to nearly 29 million°F (16 million°C), and the pressure is 100 billion times the atmospheric pressure on Earth. These extreme conditions allow nuclear fusion to take place, converting 680 million tons (620 million tonnes) of hydrogen per second into helium and energy (see p.90). This energy journeys through the radiative and convective zones to reach the visible surface.

The Sun's elements

Astronomers use spectroscopy—the close study of a spectrum—to identify chemical elements in the Sun (see pp.26–27). Atoms of these elements can be identified because they absorb or emit light of very specific colors. The Sun is so hot that some of these atoms become electrically charged plasma, causing the Sun's plasma state.

Oxygen, carbon, nitrogen, silicon, magnesium, neon, iron, and sulfur are most abundant in remaining portion

HELIUM 24%

HYDROGEN 75%

Constituent elements
The Sun's overall mass is formed of 67 elements. The majority is hydrogen and helium, the two lightest elements in the Universe.

RADIATION TAKES UP TO **1 MILLION YEARS** TO TRAVEL FROM THE **CORE** TO THE **SOLAR SURFACE**

Radiation slowly diffuses outward through radiative zone

Radiative zone is so dense that radiation only travels 0.04 in (1 mm) before encountering an obstacle

Internal structure
It takes up to 1 million years for energy from the hot, dense core to travel through the radiative and convective zones and reach the surface. The photosphere is visible from Earth, but it is covered by two layers of atmosphere, the corona and chromosphere.

CORE

Core takes up roughly inner quarter of Sun and is eight times denser than gold

PROMINENCE

Hot material follows a loop of magnetism to form a prominence

Corona, Sun's outer atmosphere, is visible during solar eclipses

Light travels through convective zone in just a few weeks

RADIATIVE ZONE

CONVECTIVE ZONE

PHOTOSPHERE

CHROMOSPHERE

CORONA

Pockets of hot gas expand, rise to surface, cool, and sink again

Once light escapes from photosphere, it reaches Earth in a little over eight minutes

Chromosphere is around 36,000°F (20,000°C)

HOW BIG IS THE SUN?

The Sun is 870,000 miles (1.4 million km) wide, and over a million Earths could fit inside it. Most stars are smaller than the Sun.

Exterior layers

The Sun's visible surface, the photosphere, is also the first layer of its atmosphere. Flamelike eruptions called prominences, and rapid energy releases known as flares, shoot up from here into the chromosphere and corona above. The corona is over 1.8 million°F (1 million°C), far hotter than the layers of atmosphere below it. This disparity in temperatures is one of the most confounding solar mysteries. Astronomers are still searching for the mechanism that injects energy into the corona, as flares alone are not sufficient.

SOLAR ECLIPSES

The Sun's faint corona is best seen during a total solar eclipse. During these spectacular events—which occur approximately every 18 months—the Moon blocks out the main glare of the Sun. Totality occurs when the Moon completely covers the main disk of the Sun. At that point, the Moon's shadow (or umbra) engulfs a portion of Earth.

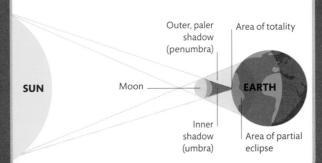

SUN

Moon

Outer, paler shadow (penumbra)

Inner shadow (umbra)

Area of totality

EARTH

Area of partial eclipse

The solar cycle

Generations of astronomers have watched solar activity rise and fall in a repeating pattern called the solar cycle. Solar activity has been scrutinized in unprecedented detail since solar telescopes were first launched into space in the 1990s.

Sunspots

The most conspicuous feature of the solar cycle is sunspots. They look like deep bruises on the Sun's surface, but they are in fact cooler regions of the photosphere at about 6,300°F (3,500°C). Magnetic fields stretch deep inside the Sun as it rotates, causing tubes of magnetism to break through the photosphere and make cup-shaped dips. Sunspots last for only a few weeks, appearing in different zones throughout the cycle.

SOLAR CONVEYOR BELTS

Giant conveyor belts of plasma churn inside the Sun's convective zone. They drag magnetic fields toward the surface and transfer material from the equator toward the poles at speeds of about 30 mph (50 kph). This causes sunspots to appear closer to the equator during the solar cycle.

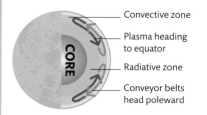

- Convective zone
- Plasma heading to equator
- Radiative zone
- Conveyor belts head poleward

CORE

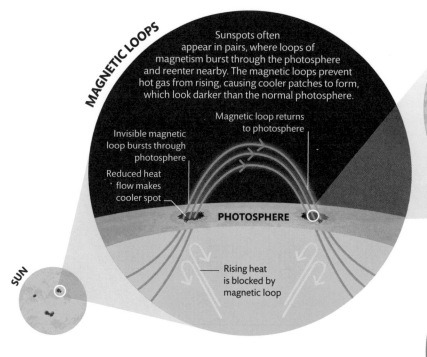

MAGNETIC LOOPS

Sunspots often appear in pairs, where loops of magnetism burst through the photosphere and reenter nearby. The magnetic loops prevent hot gas from rising, causing cooler patches to form, which look darker than the normal photosphere.

Magnetic loop returns to photosphere

Invisible magnetic loop bursts through photosphere

Reduced heat flow makes cooler spot

PHOTOSPHERE

Rising heat is blocked by magnetic loop

SUN

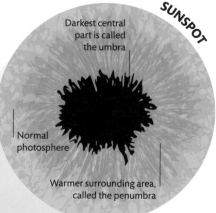

SUNSPOT

Darkest central part is called the umbra

Normal photosphere

Warmer surrounding area, called the penumbra

THE **LARGEST SUNSPOT** EVER RECORDED WAS **30** **TIMES WIDER** THAN **EARTH**

WHO DISCOVERED THE SOLAR CYCLE?

Also called the Schwabe cycle, the solar cycle was discovered in 1843 by Samuel Heinrich Schwabe, a German amateur astronomer who made daily observations over 17 years.

Solar maximum and minimum

Sunspots are not the only kind of solar activity. Gigantic eruptions called coronal mass ejections burst from the corona, and rapid releases of stored magnetic energy cause solar flares. This activity is more frequent at solar maximum and declines at solar minimum, with important consequences on Earth. Increased solar activity generates spectacular aurorae near Earth's poles (see p.31), but it can also lead to power cuts, satellite failures, and radio blackouts.

Butterfly patterns
A famous diagram called the Butterfly Diagram—because of its resemblance to the flying insect—charts the movement of the sunspot zone over the course of a solar cycle. Sunspots gradually appear closer to the equator as solar maximum nears. Comparing multiple cycles in a graph shows the variations in activity across cycles.

11-year cycle
While the average duration of the solar cycle is 11 years, over the last 400 years, it has varied in length. Recent cycles have been particularly quiet, with an unusually high number of spotless days.

YEAR 1

Cycles start with sunspots along midlatitudes

YEAR 4

Sunspots increase and appear nearer equator

YEAR 7

YEAR 10

New cycle starts as sunspots at equator wane

YEAR 12

KEY
— Previous cycle
— Current cycle
— Next cycle

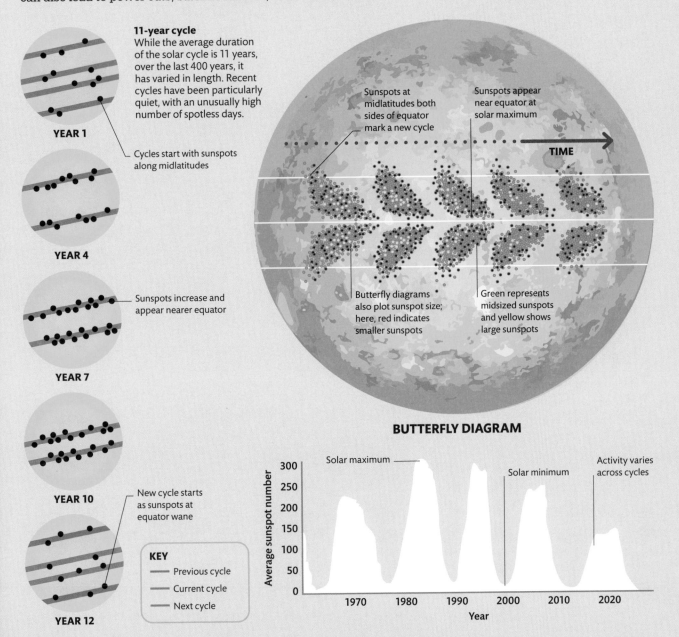

Sunspots at midlatitudes both sides of equator mark a new cycle

Sunspears appear near equator at solar maximum

TIME

Butterfly diagrams also plot sunspot size; here, red indicates smaller sunspots

Green represents midsized sunspots and yellow shows large sunspots

BUTTERFLY DIAGRAM

Solar maximum

Solar minimum

Activity varies across cycles

Average sunspot number: 300, 250, 200, 150, 100, 50, 0

Year: 1970, 1980, 1990, 2000, 2010, 2020

Earth

Called the Blue Planet, because of the expansive ocean covering 71 percent of its surface, Earth is a haven of life in space. It is the only place in the Universe unequivocally known to host living things.

Suitable for life

For life to endure on Earth, it needs to be protected from the ravages of space. Chief among these dangers is radiation from the Sun, which can damage living cells. However, Earth is enveloped in a magnetic field, arising from Earth's rotating iron core, that provides a protective shield. It helps deflect high-energy particles from the Sun and exploding stars in the wider galaxy.

Internal layers

Earth's core has remained hot since the planet first formed and continues to be heated by the decay of radioactive elements such as uranium. The temperature at the center of Earth is about 11,000°F (6,000°C), as hot as the Sun's surface. Molten material in the outer core moves and drives the magnetic field. Activity seen on the surface, such as volcanoes and earthquakes, is governed by heated material in the mantle rising through the mostly solid upper mantle and bursting through the crust.

EARTH'S CRUST HAS THE SAME RELATIVE THICKNESS AS THE SKIN ON AN APPLE

Heating the planet
The majority of the heat rising to Earth's surface is transported by convection—the same process as in the convective zone of the Sun (see pp.40–41).

Solar wind slows down and moves around Earth

Magnetosphere

Solar wind

Magnetic field stretches into a long tail

SUN

Magnetopause

Earth

Magnetosphere
The magnetosphere is a region where a magnetic field surrounds Earth. Charged particles in the solar wind slow at the surface of the magnetosphere, called the magnetopause. The field is deflected and blown into a long tail about 500 Earths wide.

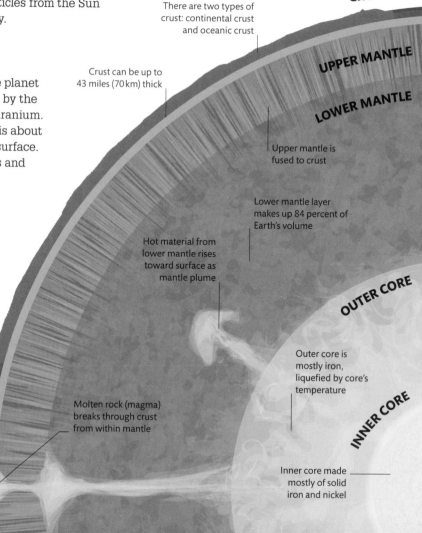

CRUST

There are two types of crust: continental crust and oceanic crust

UPPER MANTLE

LOWER MANTLE

Crust can be up to 43 miles (70 km) thick

Upper mantle is fused to crust

Lower mantle layer makes up 84 percent of Earth's volume

Hot material from lower mantle rises toward surface as mantle plume

OUTER CORE

Outer core is mostly iron, liquefied by core's temperature

INNER CORE

Molten rock (magma) breaks through crust from within mantle

Inner core made mostly of solid iron and nickel

Surface and atmosphere

Earth's crust is incredibly thin and constantly changing. It is fused to the upper mantle and broken into pieces called tectonic plates that move around on deeper parts of the mantle below. Mountains and cracks form as the plates converge or diverge. Above all this, a protective atmosphere, composed mostly of nitrogen (78 percent) and oxygen (21 percent), extends for more than 370 miles (600 km).

Tectonic plates pull away

Molten rock rises

Divergent boundary
Two tectonic plates move apart, and molten rock emerges from the mantle to fill the gap. The cooling rock forms a new piece of crust.

Plates neither collide nor pull apart

Transform boundary
Tectonic plates slide past one another, creating cracks known as faults. Most faults are found at the bottom of the ocean.

Plates slowly collide

Earth's surface changes shape

Convergent boundary
Plates collide into one another, leading to earthquakes, volcanic activity, and a deformed crust. The Himalayas formed this way.

Crust is thicker below continents than below oceans

Gases in atmosphere trap heat and help sustain life

ATMOSPHERE

WHERE DID EARTH'S WATER COME FROM?

Astronomers think water arrived on comets and asteroids, which bombarded the early Earth. These collisions left material containing water molecules deep inside Earth, from which water rose up and covered the surface.

Light rock material rose to form continents

Liquid water covered cooling Earth

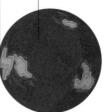

FORMING OCEAN

WHEN DID LIFE ON EARTH START?

Life on Earth is thought to have started around 4.3 billion years ago, when the planet was just half a billion years old. Before this time, the planet was too hot and lacked liquid water.

Continents and oceans are still changing shape as tectonic plates move

Protective atmosphere
Ozone, a form of oxygen in the atmosphere, protects life on Earth from ultraviolet radiation. The atmosphere also breaks up smaller asteroids and comets before they can strike the surface (see pp.28–29).

The Moon

Earth's natural satellite, the Moon, is the nearest celestial body to Earth and the most familiar object in the night sky. It is a spectacular sight when viewed through binoculars or a telescope.

How did the Moon form?

The leading idea explaining the formation of the Moon is called the giant impact hypothesis. The hypothesis suggests that within Earth's first 100 million years, it was hit by another planet of a similar size to Mars called Theia. After the impact, most of the heavy elements from both planets, such as iron and nickel, stayed on Earth to form its heavy core. At the same time, lighter, rocky material was sprayed into orbit. Gradually, gravity brought some of this debris together to form the Moon.

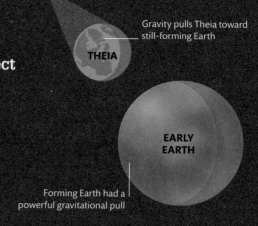

Gravity pulls Theia toward still-forming Earth

THEIA

EARLY EARTH

Forming Earth had a powerful gravitational pull

1 Collision course
Another planet—Theia—approaches the early Earth from the outer Solar System at 8,700 mph (14,000 kph).

Surface features

The distinctive lunar surface is dominated by bright areas of highland and dark patches called maria (or seas). Maria are smooth, ancient lava plains from the Moon's early volcanism, now strewn with impact craters from asteroids and comets. The mountainous highlands formed as an ocean of molten material cooled and solidified around 4.5 billion years ago. These features can be seen at their best when the Moon is partially illuminated and shadows throw the surface into sharp relief.

HOW MANY ASTRONAUTS HAVE WALKED ON THE MOON?

So far, a total of 12 astronauts have walked on the Moon. All traveled on NASA missions and stepped on the Moon between 1969 and 1972.

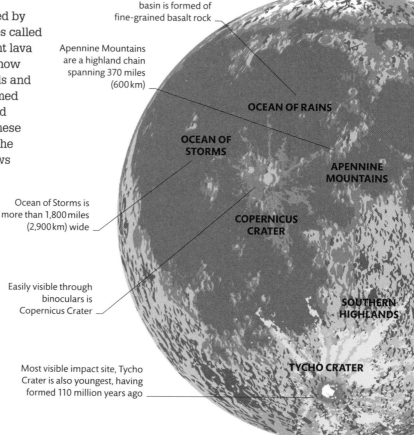

Ocean of Rains impact basin is formed of fine-grained basalt rock

Apennine Mountains are a highland chain spanning 370 miles (600 km)

OCEAN OF RAINS

OCEAN OF STORMS

APENNINE MOUNTAINS

Ocean of Storms is more than 1,800 miles (2,900 km) wide

COPERNICUS CRATER

Easily visible through binoculars is Copernicus Crater

SOUTHERN HIGHLANDS

Most visible impact site, Tycho Crater is also youngest, having formed 110 million years ago

TYCHO CRATER

Theia collides with Earth

Impact propels rocky material into space as hot vapor

Ring of debris settles around Earth

Moon's orbit follows path of debris ring

MOON

Debris forms Moon

2 **Moment of impact**
Theia collides with Earth at a 45° angle, melting rock and metal and mixing material from both worlds together.

3 **A ring forms**
Lighter material rockets into space, but much of it cannot escape Earth's gravity and settles into a ring of debris.

4 **The Moon in orbit**
Gravity pulls the ring material together into an initially molten Moon that eventually cools into the satellite it is today.

THE **MOON MOVES AWAY** FROM EARTH BY **1.5 IN (3.8 CM) EVERY YEAR**

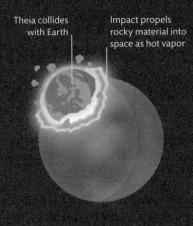

Sea of Tranquility is where Neil Armstrong first set foot in 1969

Less affected by early Earth's heat during formation, far side has fewer volcanic plains

SEA OF TRANQUILITY

Southern highlands are covered in eroded craters, implying an ancient surface

The dark side of the Moon?
Contrary to popular belief, there is no permanently "dark side" of the Moon. The back of the Moon—properly called the "far side"—is not visible from Earth but is often illuminated nonetheless.

LUNAR ECLIPSES

Lunar eclipses occur when the Moon enters the shadow of Earth. They are visible anywhere on Earth when the Moon is risen and usually appear at least twice per year. At total lunar eclipse, indirect sunlight, bent through Earth's atmosphere, turns the Moon an eerie red color. Partial lunar eclipses are also possible when the Moon moves through Earth's outer, paler shadow.

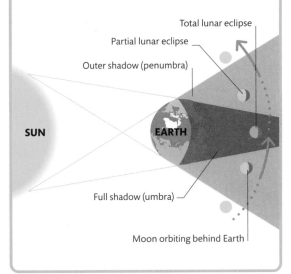

Total lunar eclipse

Partial lunar eclipse

Outer shadow (penumbra)

SUN

EARTH

Full shadow (umbra)

Moon orbiting behind Earth

Earth and the Moon

The Moon is the largest object in Earth's night sky. Its gravitational pull has slowed Earth's rotation and moves the water in our oceans, influencing our tides. Life on Earth has evolved to adapt to moonlight, tides, and the monthly lunar cycle, and the Moon is the only other world on which humans have walked.

Phases of the Moon

The Moon's changing appearance is one of the most striking features of the night sky, and its shifting shapes have been documented for millennia. Despite its apparent glow, the Moon generates no light of its own; instead, its surface reflects sunlight. Just like Earth—which at all times has one side in daylight and the other in night—the Moon is always half illuminated, but the portion that is visible from Earth changes as the Moon orbits. The lunar cycle lasts 29.5 days, slightly longer than the 27.3 days it takes for the Moon to orbit Earth. This is because Earth also moves during that time and it takes a little over two days for the Moon to realign with the Sun.

SUN

SUNLIGHT

Moon is visible during day as cycle approaches new Moon

THIRD QUARTER

Moon is waning when visible area decreases

6:00 A.M.
MERIDIAN
TRANSIT TIME

WANING CRESCENT

9:00 A.M.

NEW MOON

All sunlight falls on far side of Moon

NOON

MOON

Half of Earth is always illuminated by Sun

EARTH

WAXING CRESCENT

3:00 P.M.

FIRST QUARTER

6:00 P.M.

Moon is waxing when visible area increases

Terminator line separates light and dark

DOES THE MOON ROTATE?

The Moon rotates counterclockwise and takes as long to spin on its axis as it does to orbit Earth. This is why the same side of the Moon is always visible from Earth.

The Moon and the Sun

When the Moon is directly opposite the Sun, we can see all of the near side, creating a full Moon. When the Moon moves between Earth and the Sun, all light falls on the far side and we see a new Moon. The time that the Moon reaches its highest point in the sky (meridian transit) gradually changes through the cycle of phases.

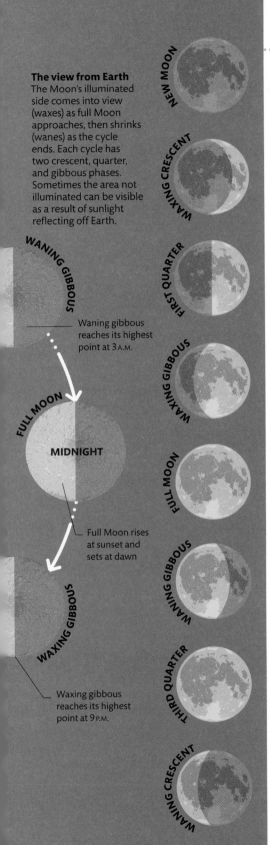

The view from Earth
The Moon's illuminated side comes into view (waxes) as full Moon approaches, then shrinks (wanes) as the cycle ends. Each cycle has two crescent, quarter, and gibbous phases. Sometimes the area not illuminated can be visible as a result of sunlight reflecting off Earth.

NEW MOON

WAXING CRESCENT

FIRST QUARTER

WAXING GIBBOUS

FULL MOON

WANING GIBBOUS

THIRD QUARTER

WANING CRESCENT

WANING GIBBOUS

Waning gibbous reaches its highest point at 3 A.M.

FULL MOON

MIDNIGHT

Full Moon rises at sunset and sets at dawn

WAXING GIBBOUS

Waxing gibbous reaches its highest point at 9 P.M.

Tides

Most places on Earth experience two high tides and two low tides every day as the planet spins through four distinct regions. The gravitational pull of the Moon causes Earth's oceans to bulge, creating high tides. When the tide goes out, the rotation of Earth is causing the tidal bulge to move away from the shore.

Tidal forces
Sea levels rise when facing the Moon as lunar gravity pulls on the water. Water is also pulled away from other areas of Earth's ocean, creating low tides. A second area of high tide on Earth's far side is due to an outward centrifugal force that exceeds the inward gravitational pull.

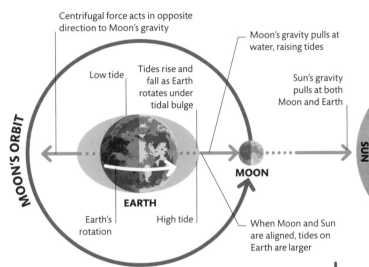

Centrifugal force acts in opposite direction to Moon's gravity

Moon's gravity pulls at water, raising tides

Low tide

Tides rise and fall as Earth rotates under tidal bulge

Sun's gravity pulls at both Moon and Earth

MOON'S ORBIT

MOON

SUN

EARTH

Earth's rotation

High tide

When Moon and Sun are aligned, tides on Earth are larger

THE MOON'S GRAVITY **LENGTHENS EARTH'S DAY** BY AN EXTRA **HALF AN HOUR EVERY 100 MILLION YEARS**

JOURNEY TO THE MOON

Six crewed spacecraft traveled a three-day flight path to the Moon between 1969 and 1972. At 43,500 miles (70,000 km) from the Moon, the spacecraft reached the neutral gravity point, where the Moon's gravity pulled the spacecraft into orbit.

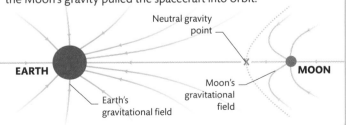

EARTH

Neutral gravity point

Earth's gravitational field

Moon's gravitational field

MOON

Mercury

The closest planet to the Sun, Mercury takes just 88 days to complete one orbit and has the most elliptical orbit of any planet. Mercury is also the smallest planet in the Solar System, with a radius of 1,500 miles (2,400 km), making it just over a third of the size of Earth.

DOES MERCURY HAVE ANY MOONS?

No. Mercury's weak gravity and proximity to the Sun meant any would-be moon material was drawn into the Sun instead.

Hollows inside craters carved out by solar winds
Smooth volcanic plains formed when early basin was flooded with lava
Mercury today has a dry, rocky surface
Streaks of material from powerful impacts surround craters
Basin is ringed by tall mountains
Volcanic plains cover 40 percent of Mercury's surface
Impact crater Munch formed 3.9 billion years ago, long after Caloris Basin
CALORIS BASIN
Craters hold material from original basin floor

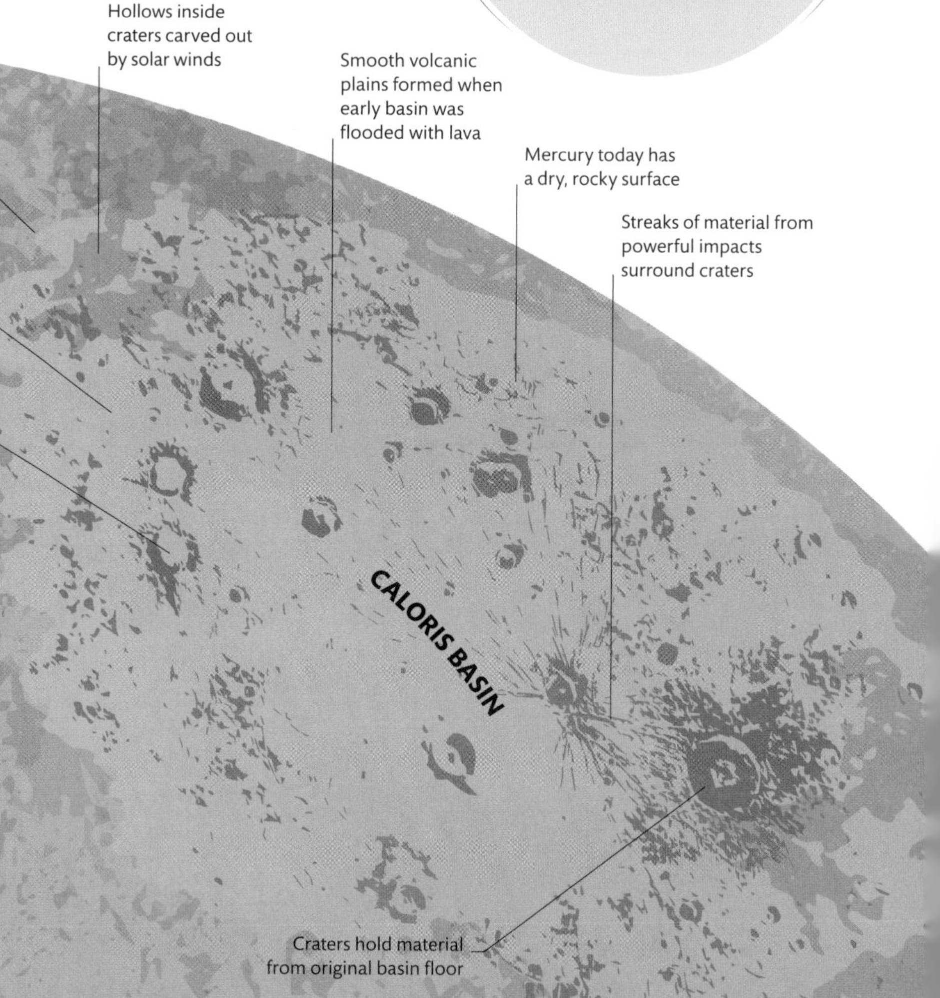

Surface features

Mercury's surface is pockmarked with countless craters. Most of these scars, from the impacts of meteoroids, date from over 4 billion years ago. They have survived almost unchanged because Mercury is too small to have any significant atmosphere. As a result, Mercury's surface greatly resembles that of the Moon. In some places, smooth plains are crisscrossed with a series of folds, caused by the whole planet gradually contracting over time.

The Caloris Basin
Mercury has one of the Solar System's largest impact basins. At over 930 miles (1,500 km) across, the Caloris Basin is around 1.5 times the width of France and is surrounded by a ring of mountains 1.2 miles (2 km) high.

MERCURY'S CRATERS ARE NAMED AFTER **ARTISTS,** INCLUDING **DISNEY, BEETHOVEN,** AND **VAN GOGH**

Atmosphere and temperature

Mercury cannot retain the significant amount of heat it receives from the Sun. During the day, the temperature climbs to over 750°F (400°C). Yet, without a thick atmosphere to trap that energy, the night side sees temperatures drop to −300°F (−180°C). This gives Mercury the biggest day-to-night temperature variation of any planet in the Solar System.

Temperature map
A map of variations in temperature below the surface of Mercury shows the hottest area (in red) directly below the Sun. This map uses observations taken with the Very Large Array (VLA) telescope in New Mexico.

KEY

750°F (400°C)   −300°F (−180°C)

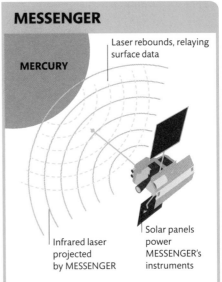

MESSENGER

MERCURY

Laser rebounds, relaying surface data

Infrared laser projected by MESSENGER

Solar panels power MESSENGER's instruments

As Mercury has no moons, NASA's MESSENGER spacecraft is probably the only object to orbit the planet in its history. Entering orbit in 2011, MESSENGER mapped 99 percent of Mercury's surface and used infrared laser signals to gather topographical data before it was deliberately crashed into the planet in 2015.

Inside Mercury

Mercury is a dense planet made up of approximately 70 percent metal and 30 percent rock—only Earth has a higher density. An iron core (which may be partly molten) takes up more than half of the planet and is surrounded by a 370-mile (600-km) wide mantle. At 20 miles (30 km) across, Mercury's rocky crust has a similar thickness to Earth's.

Space mission data
Data collected through space missions, including Mariner 10 and MESSENGER, has informed astronomers of Mercury's internal layers. MESSENGER also found evidence of water ice at Mercury's poles.

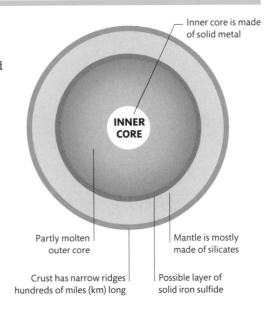

Inner core is made of solid metal

INNER CORE

Partly molten outer core

Crust has narrow ridges hundreds of miles (km) long

Mantle is mostly made of silicates

Possible layer of solid iron sulfide

Venus

The second plant from the Sun is often referred to as Earth's twin, as it is only slightly smaller than Earth and has several familiar features, including mountains and volcanoes. However, Venus also has some unique structures.

Surface features

The giant volcano known as Maat Mons towers 5 miles (8 km) above the surface of Venus. No other planet has more volcanoes, meaning that the Venusian surface is strewn with evidence of ancient lava flows and intense volcanic activity. Distinctive volcanic domes that resemble pancakes are scattered in clusters across the planet, as are deep impact craters from large meteorites. Raised circular or oval structures hundreds of miles across also litter the surface. Called coronas, they were caused by hot magma welling up into the crust.

WHY DOES VENUS LOOK SO BRIGHT?

Venus appears to be bright when viewed from Earth, because its atmosphere is filled with thick sulfuric acid clouds. Sunlight is reflected off these clouds, making it appear to shine.

A **DAY ON VENUS—** THE TIME FROM ONE SUNRISE TO THE NEXT— LASTS **117 EARTH DAYS**

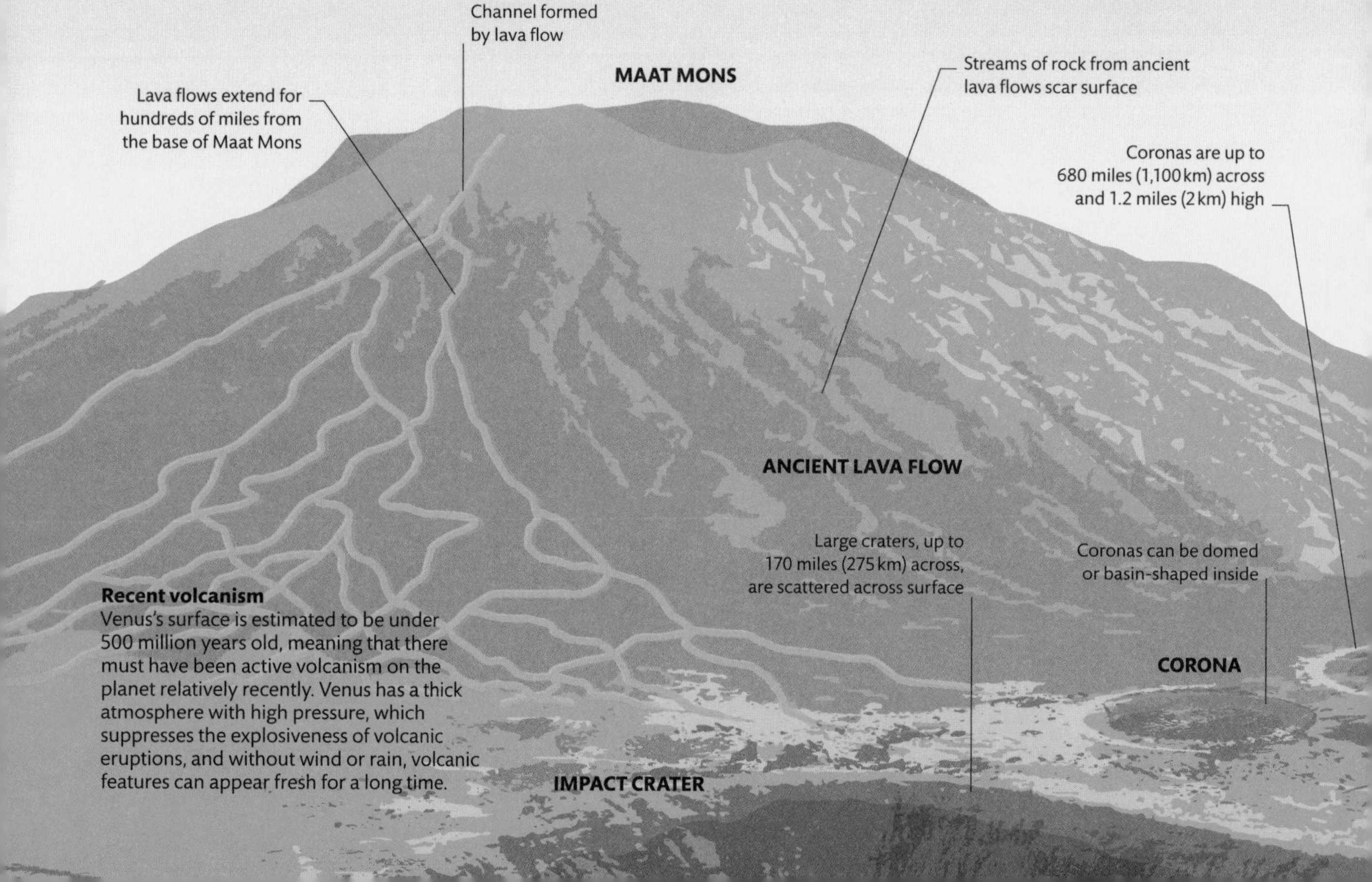

Channel formed by lava flow

MAAT MONS

Lava flows extend for hundreds of miles from the base of Maat Mons

Streams of rock from ancient lava flows scar surface

Coronas are up to 680 miles (1,100 km) across and 1.2 miles (2 km) high

ANCIENT LAVA FLOW

Large craters, up to 170 miles (275 km) across, are scattered across surface

Coronas can be domed or basin-shaped inside

CORONA

Recent volcanism

Venus's surface is estimated to be under 500 million years old, meaning that there must have been active volcanism on the planet relatively recently. Venus has a thick atmosphere with high pressure, which suppresses the explosiveness of volcanic eruptions, and without wind or rain, volcanic features can appear fresh for a long time.

IMPACT CRATER

TRANSITS OF VENUS

Venus passes directly between Earth and the Sun in a rare event called a transit. Two transits of Venus occur over an eight-year period, but then it is more than a century before the next pair occur. The next transits will be in 2117 and 2125. The time it takes for Venus to transit the Sun was initially used to calculate the Earth–Sun distance, and transits are still invaluable to astronomers. Sunlight to Earth dims slightly during transits, and astronomers look for similar occurrences to identify Earth-sized planets orbiting nearby stars.

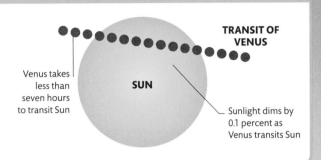

TRANSIT OF VENUS

Venus takes less than seven hours to transit Sun

SUN

Sunlight dims by 0.1 percent as Venus transits Sun

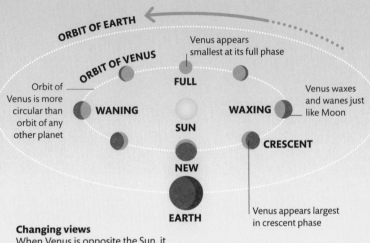

ORBIT OF EARTH

ORBIT OF VENUS

Orbit of Venus is more circular than orbit of any other planet

Venus appears smallest at its full phase

FULL

WANING

SUN

Venus waxes and wanes just like Moon

WAXING

CRESCENT

NEW

EARTH

Venus appears largest in crescent phase

Changing views
When Venus is opposite the Sun, it appears on Earth to be fully lit. When it is closest to Earth, most sunlight falls onto the far side of Venus, revealing only a sliver of the planet.

Phases of Venus

Italian astronomer Galileo Galilei spotted in 1610 that, like the Moon, Venus has phases, thus proving that all planets—including Earth—circle the Sun. As Venus orbits the Sun, its illumination viewed from Earth appears to change. Slivers of Venus appear to be bigger and brighter as it nears Earth, then as Venus passes behind the Sun, a full hemisphere is visible. The cycle takes over two-and-a-half Venusian years (584 Earth days) to complete.

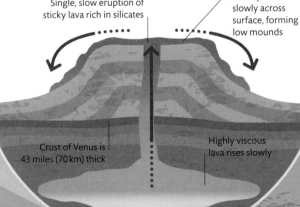

PANCAKE DOME FORMATION

Flat volcanic structures, known as pancake domes, are unique to Venus. Thick, slow-moving lava rises through a central vent and spreads up to 100 times farther than it does from similar structures on Earth.

Single, slow eruption of sticky lava rich in silicates

Lava spreads slowly across surface, forming low mounds

Crust of Venus is 43 miles (70 km) thick

Highly viscous lava rises slowly

PANCAKE DOME

Greenhouse effect
Certain gases, such as carbon dioxide, water vapor, and methane, act like the glass in a greenhouse—they let solar energy through but prevent it from escaping. The result is a sharp rise in temperature.

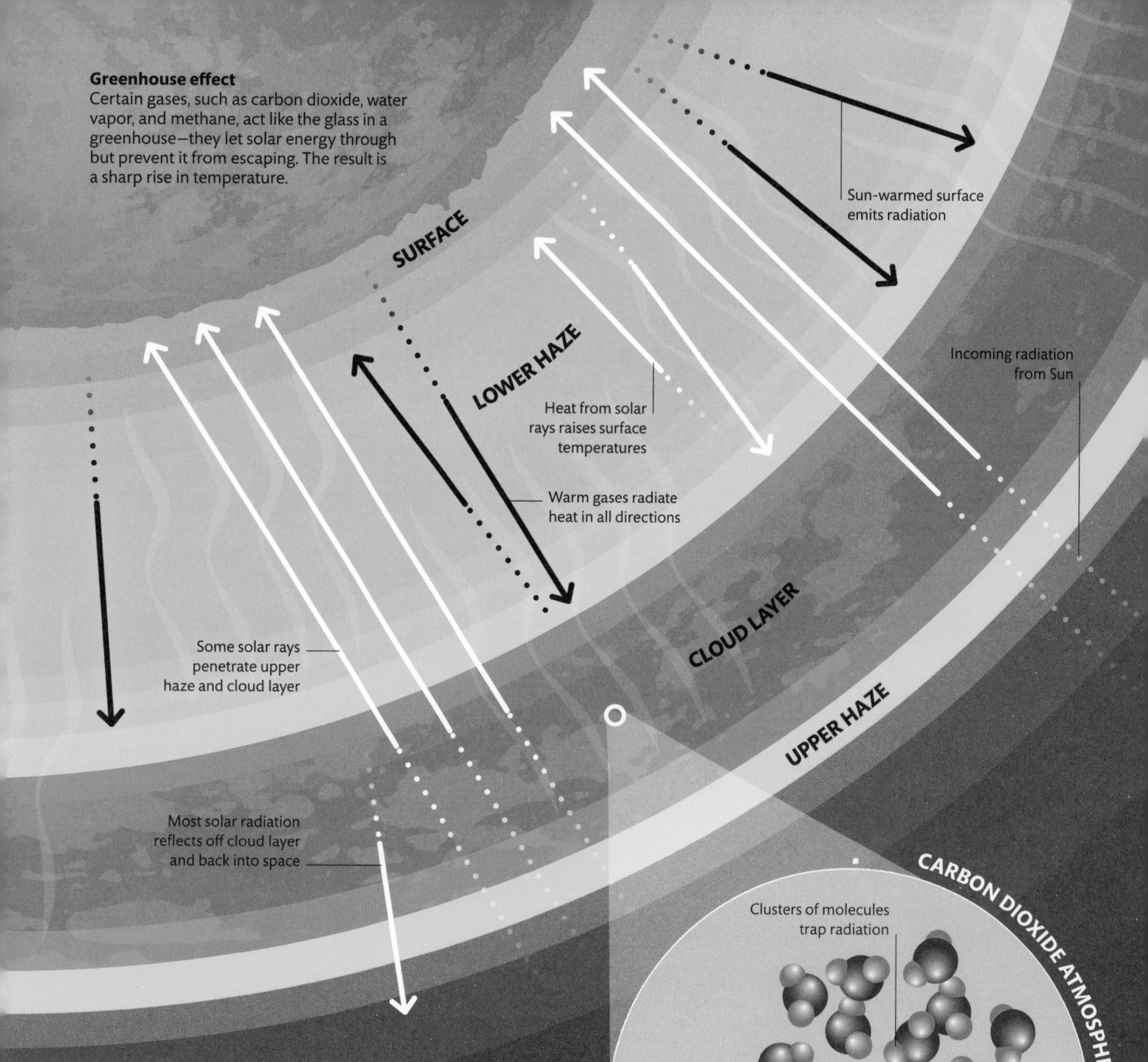

SURFACE

LOWER HAZE

Sun-warmed surface emits radiation

Incoming radiation from Sun

Heat from solar rays raises surface temperatures

Warm gases radiate heat in all directions

CLOUD LAYER

UPPER HAZE

Some solar rays penetrate upper haze and cloud layer

Most solar radiation reflects off cloud layer and back into space

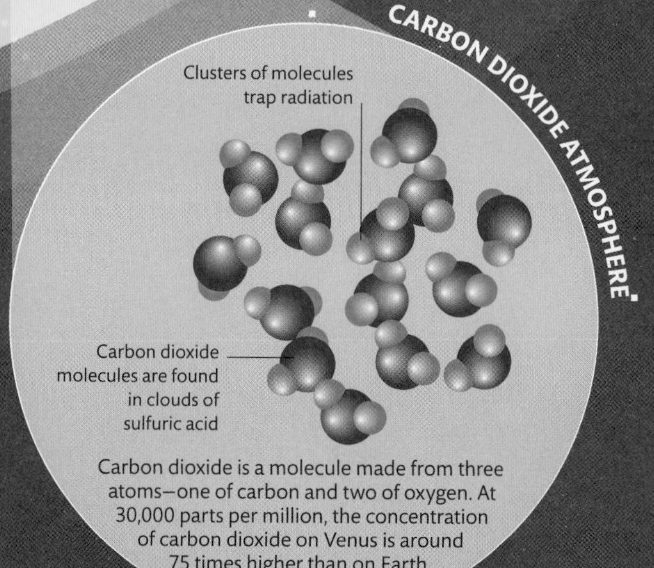

CARBON DIOXIDE ATMOSPHERE

Clusters of molecules trap radiation

Carbon dioxide molecules are found in clouds of sulfuric acid

Carbon dioxide is a molecule made from three atoms—one of carbon and two of oxygen. At 30,000 parts per million, the concentration of carbon dioxide on Venus is around 75 times higher than on Earth.

Runaway greenhouse effect

As we know from recent climate change on Earth, carbon dioxide has a potent warming effect. Water vapor is a powerful greenhouse gas, too. Carbon dioxide and water vapor released by volcanic activity built up in Venus's atmosphere, which got hotter and hotter. As the water broke down or escaped, more carbon dioxide formed, warming the planet even further. Once this process had started, it could not stop and became a runaway effect.

Hothouse planet

Our nearest planetary neighbor, Venus is the hottest planet in the Solar System and a sweltering greenhouselike world with an extreme climate.

Super-rotation

A quirk of Venus is that the time it takes to rotate relative to the stars is longer than its year of 225 Earth days. It also rotates in the opposite direction to the other planets. One full rotation takes 243 Earth days, although the solar day on Venus is shorter, lasting 117 days. Despite Venus's sluggish rotation, high-speed winds whip around the upper atmosphere over the equatorial region in just four days. This super-rotation is partly due to heat from the Sun causing variations in the atmospheric pressure, but the causes are not fully understood.

STANDING ON VENUS WOULD FEEL LIKE HAVING 15 ELEPHANTS ON YOUR SHOULDERS

Cloud top

Venus's rotation

Convection cell circulation

Equator

Wind direction

Planet's surface

Polar collar (body of colder gas)

Inner circulation

Hot gas rises at the equator and flows toward the poles, where it cools and sinks back down to be reheated. These conveyor belts of gas circulating across Venus are called convection cells.

IS THERE LIFE ON VENUS?

There may be life on Venus, although there is no current evidence. Some scientists argue that life could persist in the cooler regions of the upper atmosphere.

DID VENUS HAVE WATER?

Venus may not have always been such a hostile environment. Billions of years ago, before the greenhouse effect, the planet may have been more Earth-like. Infrared mapping reveals lower-lying regions that may have contained shallow oceans.

Hotter areas show low altitudes of potential oceans

Slightly cooler, high altitudes may have been ancient continents

SOUTHERN HEMISPHERE

Structure and composition

With the exception of oceans, the surface of Mars has many similarities to the surface of Earth. Soaring mountain ranges; ice caps; towering volcanoes; and long, deep valley systems can all be seen on Mars. Far below the surface sits a core that is roughly 1,300 miles (2,100 km) in radius and is formed of mostly iron and nickel, although there is a small amount of sulfur, too. There is also evidence from magnetized areas of the crust that Mars once had a magnetic field, but without a molten core, it faded away.

The surface of Mars
Mars's surface is strikingly varied, with nonvolcanic low-lying regions dominating the surface north of the equator and highlands and extinct volcanoes concentrated in the southern hemisphere.

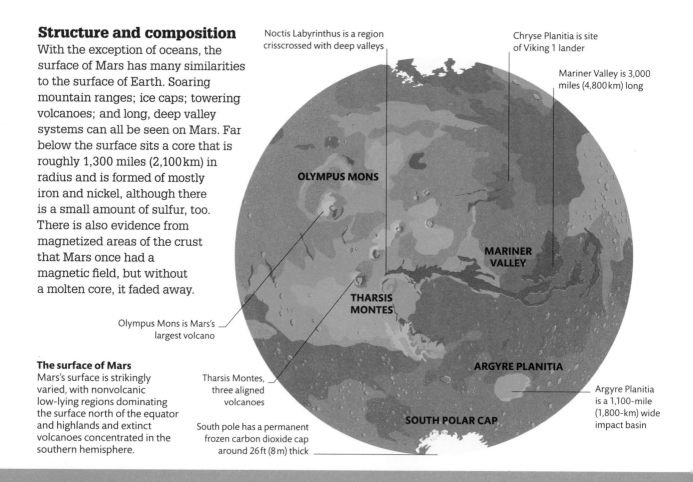

Noctis Labyrinthus is a region crisscrossed with deep valleys

Chryse Planitia is site of Viking 1 lander

Mariner Valley is 3,000 miles (4,800 km) long

OLYMPUS MONS

MARINER VALLEY

THARSIS MONTES

Olympus Mons is Mars's largest volcano

Tharsis Montes, three aligned volcanoes

South pole has a permanent frozen carbon dioxide cap around 26 ft (8 m) thick

ARGYRE PLANITIA

SOUTH POLAR CAP

Argyre Planitia is a 1,100-mile (1,800-km) wide impact basin

Mars

No other planet has captured the human imagination quite like Mars, the fourth planet from the Sun. The Red Planet continues to attract daring rover missions to explore its desertlike surface.

Internal structure
Around Mars's dense core is a thick, rocky mantle; a crust; and a thin atmosphere of carbon dioxide, nitrogen, and argon. Mars is still seismically active and has hundreds of "Marsquakes" each year.

Thin atmosphere provides little protection

Thin crust of dust-covered volcanic rock

Mantle of silicate rock

Dense core may be partly liquid

MARS'S MOONS

Mars has two moons that are much smaller than the major satellites of other planets. The moons could have formed from material thrown into Mars's orbit by impacts or have once been asteroids from the neighboring Main Belt.

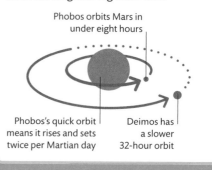

Phobos orbits Mars in under eight hours

Phobos's quick orbit means it rises and sets twice per Martian day

Deimos has a slower 32-hour orbit

Syrtis Major Planum, a low shield volcano, is so prominent, it was first permanent feature seen from Earth

North polar ice cap is 600 miles (1,000 km) wide in summer

NORTH POLAR CAP

SYRTIS MAJOR PLANUM

HELLAS BASIN

Hellas Basin is a large, round impact crater over 4 miles (7 km) deep

MARS'S RED COLOR COMES FROM IRON OXIDE (RUST)

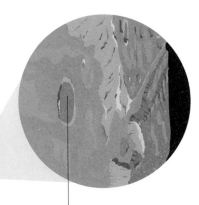

Gusev Crater once held water or water ice from a nearby channel

Spirit landing site
In 2004, NASA's Spirit rover landed on an ancient lake bed called Gusev Crater. Spirit spent 1,944 days exploring the area before it became stuck in the crater's soft sand.

The search for life

Of all the planets in the Solar System, Mars is most likely to have supported life in its past. It is thought that the Red Planet had a much wetter past, with oceans and lakes sprawled across its surface and ancient rivers meandering across the Martian terrain. As every living thing on Earth requires liquid water to survive, its presence on Mars suggests life might also have gained a foothold when the climate was more favorable. Scientists are searching for signs of biological activity and even question whether life could exist on Mars in the future.

Mars Odyssey holds the record for the longest continuous service in Mars orbit

ExoMars is studying methane in search of life

1971 Mars 2
1971 Mars 3
1971 Mariner 9
1975 Viking 1
1975 Viking 2
1996 Mars Global Surveyor
2001 Mars Odyssey
2003 Mars Express
2005 Mars Reconnaissance Orbiter
2013 Mars Orbiter Mission
2013 MAVEN
2016 ExoMars Trace Gas Orbiter

Successful orbiters
More spacecraft have successfully orbited Mars than any other planet. Their missions have included detailed mapping and communication with rovers and other probes on the surface.

Martian ice and volcanoes

Two of the most conspicuous features on the surface of Mars are its ice caps and volcanoes. Together, they hold many secrets of Mars's past and have been heavily scrutinized by scientists.

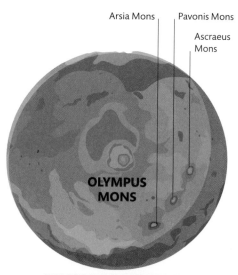

Arsia Mons
Pavonis Mons
Ascraeus Mons

OLYMPUS MONS

THARSIS BULGE FROM ABOVE

Volcanoes

One region of Mars is synonymous with volcanoes—the Tharsis Bulge. Straddling the Martian equator to the west of the Mariner Valley, the Tharsis Bulge is a volcanic plateau formed by the upwelling of more than a billion billion tons (tonnes) of material from inside Mars. It is so massive, it may have affected the tilt of Mars's rotation axis. On or close to the bulge sit four large volcanoes, including the colossal Olympus Mons, all of which are taller than Mount Everest on Earth.

Olympus Mons

Mars's tallest summit is also the Solar System's highest volcanic peak. Olympus Mons is so sprawling, it covers an area of 116,000 square miles (300,000 square km), approximately the same size as Italy. It is also relatively shallow, with an average incline of just 5°.

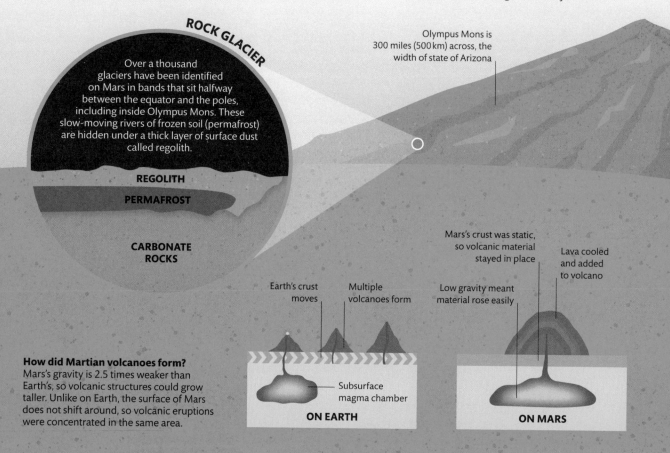

ROCK GLACIER

Over a thousand glaciers have been identified on Mars in bands that sit halfway between the equator and the poles, including inside Olympus Mons. These slow-moving rivers of frozen soil (permafrost) are hidden under a thick layer of surface dust called regolith.

REGOLITH

PERMAFROST

CARBONATE ROCKS

Olympus Mons is 300 miles (500 km) across, the width of state of Arizona

Earth's crust moves

Multiple volcanoes form

Subsurface magma chamber

ON EARTH

Mars's crust was static, so volcanic material stayed in place

Lava cooled and added to volcano

Low gravity meant material rose easily

ON MARS

How did Martian volcanoes form?

Mars's gravity is 2.5 times weaker than Earth's, so volcanic structures could grow taller. Unlike on Earth, the surface of Mars does not shift around, so volcanic eruptions were concentrated in the same area.

Water and ice

Mars is bookended by two vast polar ice caps, which grow and shrink with the seasons and are around 2 miles (3 km) thick. If all the ice melted, the liquid would flood Mars to a depth of over 16 ft (5 m). The ice caps contain water and frozen carbon dioxide that turns into a gas at warmer temperatures. This seasonal release of gas causes fierce winds to blow dust around the planet. Ice has also been spotted beneath the surface farther from the poles, scuffed up by the wheels of trundling Mars rovers.

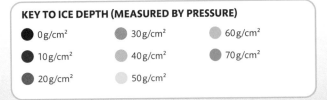

High concentrations of carbon dioxide ice

Carbon dioxide ice recedes as Mars warms

EARLY SPRING

LATE SPRING

KEY TO ICE DEPTH (MEASURED BY PRESSURE)

- $0\,g/cm^2$
- $10\,g/cm^2$
- $20\,g/cm^2$
- $30\,g/cm^2$
- $40\,g/cm^2$
- $50\,g/cm^2$
- $60\,g/cm^2$
- $70\,g/cm^2$

THE **WATER** IN **MARTIAN ICE** COULD **COVER** THE **PLANET** WITH **OCEANS** 115 FT (35 M) DEEP

OLYMPUS MONS

Mount Everest is only about one-third as tall as Olympus Mons

MOUNT EVEREST

ARE MARS'S VOLCANOES STILL ACTIVE?

Most scientists think not, but some argue that the volcanoes are dormant. Liquid water found deep under the surface might have been thawed by magma chambers.

THE MARINER VALLEY

Mariner Valley 2,500 miles (4,000 km)

USA coast to coast 2,800 miles (4,500 km)

USA

Grand Canyon 277 miles (446 km)

MARINER VALLEY

Mariner Valley 5 miles (8 km)

Grand Canyon 1 mile (1.6 km)

At over 2,500 miles (4,000 km) long and 5 miles (8 km) deep, the gigantic and intricate system of the Mariner Valley cuts a quarter of the way around the Martian equator. The huge volcanic crack in Mars's crust formed 3.5 billion years ago as the planet cooled. It is named after the Mariner 9 spacecraft, which spotted it while orbiting the Red Planet in the early 1970s.

Asteroids

There is more to the Solar System than the Sun, its planets, and their moons. Small lumps of rock and metal called asteroids are littered between the planets in orbits around the Sun.

Asteroids and the early Solar System

Asteroids appear in the sky as starlike specks of light, but they are in fact rocky and metallic objects orbiting the Sun. They are the leftover building blocks of the Solar System and as such predate the planets. That makes asteroids invaluable tools for understanding the formation of the Solar System. The meteorites that periodically land on Earth are mostly fragments of asteroids. By analyzing their radioactive impurities, scientists can estimate their age and, in turn, the age of the Solar System.

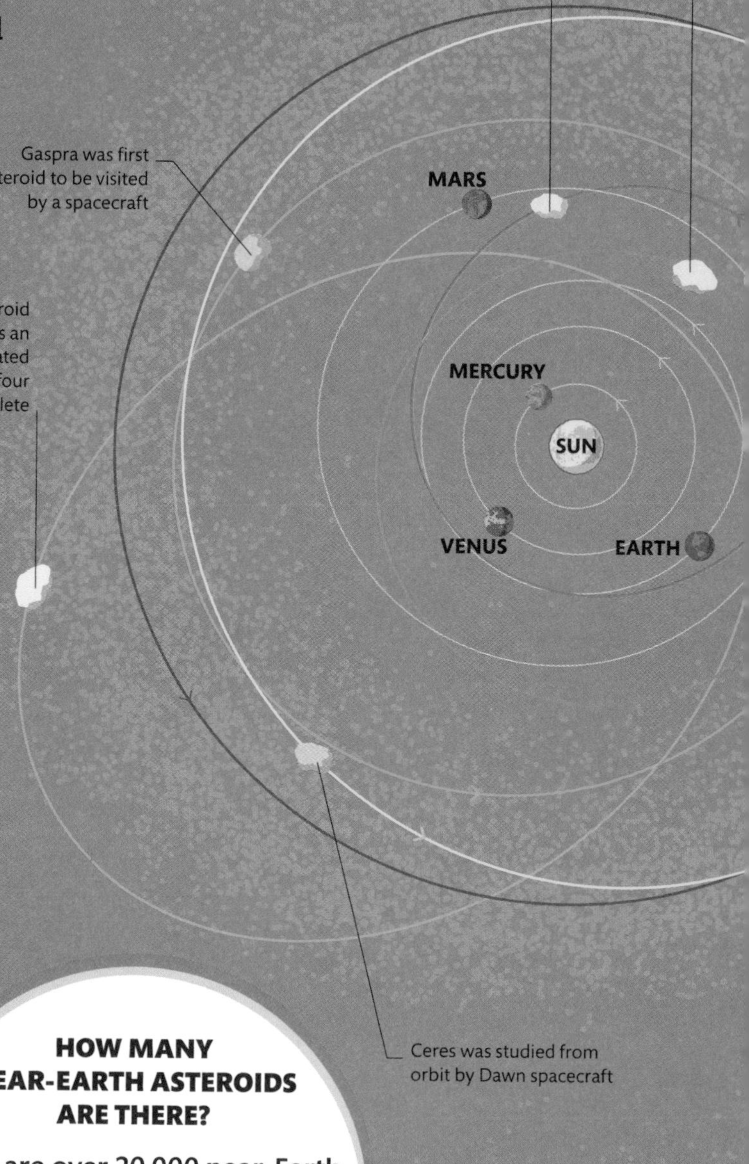

Eros, first near-Earth asteroid discovered, has a short orbit of less than two years

Itokawa orbits very close to Earth every two years

Gaspra was first asteroid to be visited by a spacecraft

Near-Earth asteroid Toutatis has an unusual elongated orbit that takes four years to complete

MARS

MERCURY

SUN

VENUS

EARTH

Ceres was studied from orbit by Dawn spacecraft

TROJAN ASTEROIDS

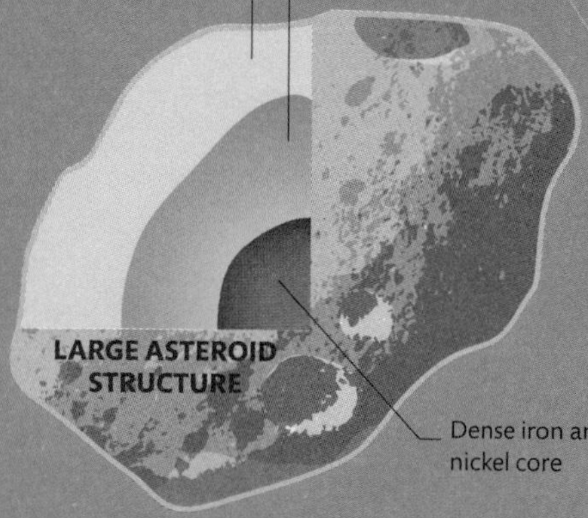

Outer layer of silicate material

Layer of iron and nickel mixed with silicates

LARGE ASTEROID STRUCTURE

Dense iron and nickel core

What is an asteroid?
Formed of materials such as silicates, nickel, and iron fused together and impacted by collisions, asteroids are small orbiting bodies. The largest asteroid, Ceres, is almost 590 miles (950 km) across and is also classed as a dwarf planet.

HOW MANY NEAR-EARTH ASTEROIDS ARE THERE?

There are over 20,000 near-Earth asteroids known to move in the vicinity of our planet. Scientists are developing ways to stop any potentially dangerous collisions with Earth.

Asteroids in the Solar System

Ninety percent of asteroids are found in the Main Belt, also known as the Asteroid Belt, between the orbits of Mars and Jupiter. Smaller clusters of asteroids called Trojan asteroids trail Jupiter's path around the Sun, trapped by the giant planet's gravity. Many asteroids, known as near-Earth asteroids, also orbit closer to Earth. Some of these cross Earth's orbit and could potentially collide with our planet.

JUPITER

MAIN BELT

Orbit of Ida, the first asteroid found to have a moon, crosses the path of Ceres

ASTEROID TYPES
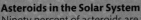

Types of asteroid

There are three main types of asteroid, grouped by their characteristics.

 Si Silicon   **Fe** Iron   **Mg** Magnesium

S-Type
This moderately bright type is made of silicate rocks and metals, with hardly any water.

C Carbon   **P** Phosphorus   **N** Nitrogen

C-Type
A very dark type made of rocks and clay minerals, with high carbon content and hardly any metals.

 Fe Iron   **Ni** Nickel

M-Type
A moderately bright type with high metal content, made of rock and water containing minerals.

Extinction-level events

Asteroids that collide with Earth can cause death and destruction. Sixty-six million years ago, an asteroid the size of a small city, the Chicxulub Impactor asteroid, careened into the coast of Mexico at Chicxulub, triggering an apocalyptic event that wiped the dinosaurs from the world. Similarly sized events strike approximately every 100 million years.

Asteroid sizes

The asteroid that hastened the demise of the dinosaurs was wider than Mount Everest is tall. But, it was small compared to the largest asteroids, which are over 300 miles (500 km) wide.

↑
330 miles (530 km)
↓

5.5 miles
(8.9 km)

6 miles
(10 km)

**MOUNT
EVEREST**

**CHICXULUB IMPACTOR
(ASTEROID)**

**VESTA
(ASTEROID)**

THE COMBINED MASS OF ALL THE ASTEROIDS IS JUST 3 PERCENT OF THE MASS OF THE MOON

GRABBING AN ASTEROID

Rather than waiting for meteorites to deliver asteroid samples to Earth, the Japanese space agency (JAXA) dispatched the Hayabusa probe to land on the asteroid Itokawa in 2005. It grabbed 1,500 dust particles to inform our understanding of the asteroid's formation before returning to Earth and landing in the Australian Outback.

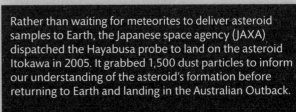
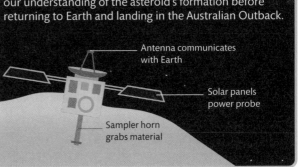

Antenna communicates with Earth

Solar panels power probe

Sampler horn grabs material

How Vesta formed

Asteroids, also known as minor planets, are leftover building blocks of planets. The planets started to grow when gravity attracted small pieces of material together to make chunks called planetesimals. Not all of the pieces were incorporated into planets, and a belt was left between Mars and Jupiter. However, some of the most massive, such as Vesta, grew hot enough to melt and were rounded by their own gravity. Smaller planetesimals kept their irregular shapes.

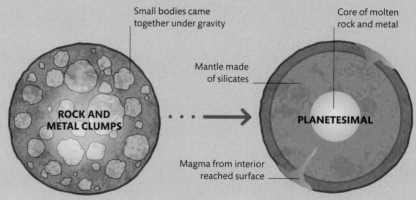

Small bodies came together under gravity

ROCK AND METAL CLUMPS

Core of molten rock and metal

Mantle made of silicates

PLANETESIMAL

Magma from interior reached surface

1 **Aggregation of small bodies**
Gravity drew lumps of rock and metal together, causing them to collide. The material formed a planetesimal, and the energy of the impacts caused melting.

2 **Heavy elements sink**
A lump of molten rock and metal formed. The heaviest elements—such as iron and nickel—sank to the center to create a core, and magma flowed to the surface.

Exploring asteroids

To learn more about asteroids and the Main Belt, scientists study them with instruments, such as the Hubble Space Telescope, and dispatch spacecraft, such as NASA's Dawn, to make detailed observations and return material to Earth.

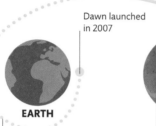

Dawn launched in 2007

Distinctively shaped Snowman Craters from collisions with other asteroids

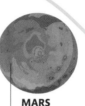

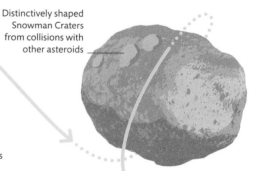

EARTH

Gravitational slingshot from Earth boosted Dawn's speed

MARS

Dawn made fly-by observations of Mars

VESTA

Differing asteroids

Ceres and Vesta are neighbors in the Main Belt, but they are not alike. Vesta is the smaller of the two, at 355 miles (570 km) across to Ceres's 590 miles (950 km). Vesta is also closer to the Sun, and it is dense and rocky like the terrestrial planets. In fact, it is thought Earth was made from bodies like Vesta colliding. Ceres's additional distance from the Sun means it is cold enough to retain water ice, making its structure more like some of the icy moons of the outer Solar System.

Dawn reached top speed of 25,000 mph (41,000 kph)

Ceres and Vesta

There are over a million asteroids in the Main Belt (see pp.60–61), but just two account for 40 percent of their combined mass—Ceres, which is also classified as a dwarf planet, and Vesta.

COULD THERE BE LIFE ON CERES?

Ceres is a good place to search for potential signs of life. It has water and possibly a hot core. However, if there are any signs of life, it is likely to have been in Ceres's distant past.

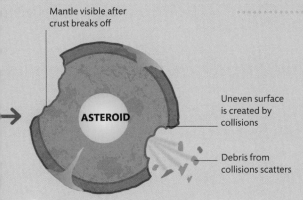

Mantle visible after crust breaks off

ASTEROID

Uneven surface is created by collisions

Debris from collisions scatters

3 Impact breaks off fragments
Later collisions chipped away at the solidified surface, further contributing to an uneven shape. Particularly big impacts exposed deep-lying inner layers.

WHITE SPOTS ON CERES

As NASA's Dawn approached Ceres in 2015, it saw bright spots on the floor of the Occator Crater. They appear to be highly reflective salty deposits, possibly left behind when water evaporated away from Ceres and into space. Astronomers suspect there is a deep reservoir of salty water inside Ceres that periodically reaches the surface.

White spots apparent on surface

CERES

NASA's Dawn mission

NASA's Dawn mission studied Ceres and Vesta in order to reveal clues about the beginning of the Solar System. Instruments on board were designed to work out the asteroids' compositions and help explain the evolutionary paths that made them so different. The mission also demonstrated the power of an ion engine (see pp.192–193).

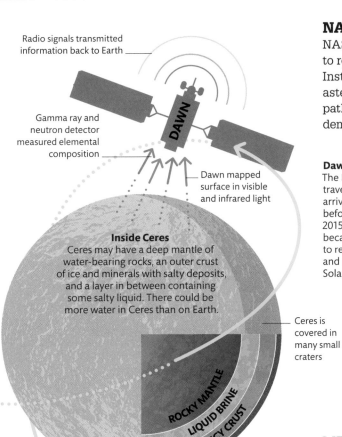

Radio signals transmitted information back to Earth

Gamma ray and neutron detector measured elemental composition

Dawn mapped surface in visible and infrared light

DAWN

Inside Ceres
Ceres may have a deep mantle of water-bearing rocks, an outer crust of ice and minerals with salty deposits, and a layer in between containing some salty liquid. There could be more water in Ceres than on Earth.

Ceres is covered in many small craters

ROCKY MANTLE
LIQUID BRINE
ICY CRUST

CERES

Dawn's flight path
The Dawn spacecraft traveled via Mars to arrive at Vesta in 2011, before reaching Ceres in 2015. In doing so, it became the first mission to reach the Main Belt and orbit two different Solar System bodies.

JUPITER

Dawn launched from Earth in September 2007

Dawn departs Vesta

SUN
EARTH

MAIN BELT

Mars provided a boost for Dawn

MARS
VESTA

CERES

Began orbiting Ceres in February 2015

Mission ended in July 2015

In July 2011, Dawn began to orbit Vesta, where it stayed for over a year

VESTA'S RHEASILVIA CRATER CONTAINS THE TALLEST MOUNTAIN IN THE SOLAR SYSTEM

Jupiter

Jupiter is so big that all of the other planets in the Solar System could fit inside it. This gas giant, with its strong gravitational pull, dominates everything around it.

Internal layers

Jupiter has a radius of almost 43 million miles (70 million km), and its gargantuan size puts its internal layers under extreme pressure from the weight of material above. The planet is mostly made of hydrogen and helium. In the outer layer, these elements are gases, but deeper inside Jupiter, the gases are gradually crushed and become liquid. At around 12,000 miles (20,000 km) deep, they become an electrically charged liquid called metallic hydrogen. This layer forms the largest ocean in the Solar System. Beneath it is probably a hot core with a temperature of around 90,000°F (50,000°C).

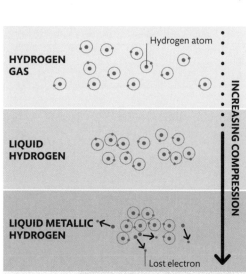

HYDROGEN GAS

Hydrogen atom

LIQUID HYDROGEN

LIQUID METALLIC HYDROGEN

Lost electron

INCREASING COMPRESSION

Compressed layers
As pressure increases, hydrogen atoms are pressed together, becoming liquid and eventually losing electrons. This makes the liquid electrically charged and metallic, which means it can conduct electric currents and generate magnetic fields.

Auroral ovals
Electric energy at Jupiter's poles causes auroral ovals 600 miles (1,000 km) wide. In these ovals, bright spots appear where Jupiter's magnetosphere draws charged particles from nearby moons.

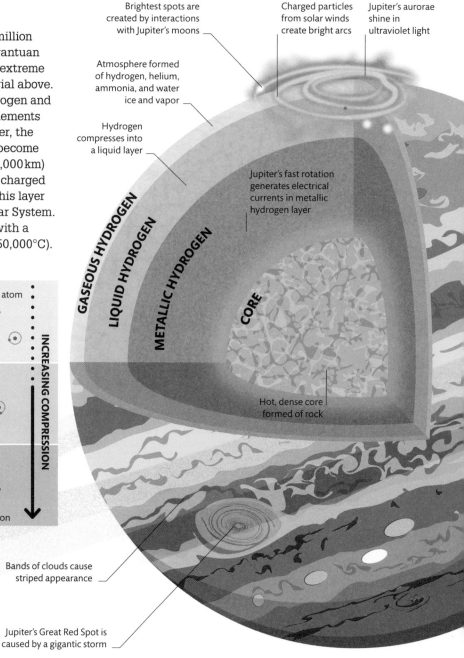

Brightest spots are created by interactions with Jupiter's moons

Charged particles from solar winds create bright arcs

Jupiter's aurorae shine in ultraviolet light

Atmosphere formed of hydrogen, helium, ammonia, and water ice and vapor

Hydrogen compresses into a liquid layer

Jupiter's fast rotation generates electrical currents in metallic hydrogen layer

GASEOUS HYDROGEN

LIQUID HYDROGEN

METALLIC HYDROGEN

CORE

Hot, dense core formed of rock

Bands of clouds cause striped appearance

Jupiter's Great Red Spot is caused by a gigantic storm

DOES JUPITER HAVE RINGS?

Yes, like the other three giant planets, Jupiter has rings. The rings are made from dust and are hard to see from Earth. They were spotted in 1979 by the Voyager 1 spacecraft.

HOT-JUPITERS

Astronomers have found many Jupiter-sized exoplanets close to other stars. These hot-Jupiters (see pp.102–103) orbit their host stars in under 10 days. It is thought that they formed farther away from their host stars and migrated inward over time, possibly pulled by the gravity of a companion star orbiting the host star.

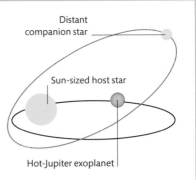

Distant companion star

Sun-sized host star

Hot-Jupiter exoplanet

Jupiter is surrounded by four rings

Rings are formed of small, dark dust particles

RINGS

Giant planet
Jupiter is so large that Earth could fit inside it more than 1,000 times. It has a bulge at the equator with flattened poles, caused by Jupiter's swift rotation.

JUPITER HAS THE **SHORTEST DAY** IN THE SOLAR SYSTEM AT **9 HOURS AND 56 MINUTES**

Magnetosphere

Jupiter has a magnetic field so big, it extends up to 2 million miles (3 million km) toward the Sun, and the magnetotail behind Jupiter is more than 600 million miles (1 billion km) long, stretching out beyond Saturn's orbit. The magnetosphere's colossal size is a result of the huge convective currents generated inside Jupiter's subsurface ocean of metallic hydrogen.

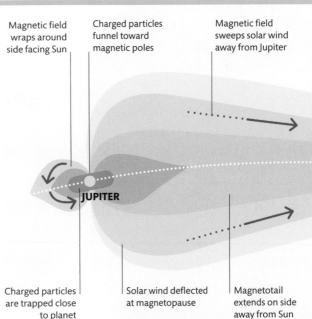

Magnetic field wraps around side facing Sun

Charged particles funnel toward magnetic poles

Magnetic field sweeps solar wind away from Jupiter

JUPITER

Clouds mainly made of ammonia ice

Powerful field
Jupiter's magnetic field is up to 54 times more powerful than Earth's magnetic field. It traps charged particles and accelerates their movement to incredibly high speeds.

Charged particles are trapped close to planet

Solar wind deflected at magnetopause

Magnetotail extends on side away from Sun

The Great Red Spot

The giant oval storm in Jupiter's southern hemisphere, the Great Red Spot, is the planet's most distinctive feature. It is a colossal anticyclone and the biggest storm in the Solar System. It has been observed since at least the 1830s, and in that time it has halved in size, although it is not known why. It is now about the same size as Earth and could become circular by 2040.

Storms on Jupiter

White oval storms are some of the most common types of storm seen on Jupiter. In December 2019, NASA's Juno spacecraft watched as two ovals merged together over several days.

Near north pole, a large cold spot is linked to Jupiter's aurorae

Row of white spots known as String of Pearls

HEATED ATMOSPHERE

Energy release

The Great Red Spot consists of rotating clouds with eddies joining at its edges. The region above the spot is hotter than any other part of Jupiter's atmosphere. It is thought that this is due to the storm compressing and heating gases. The heat energy is then transferred upward.

Rising energy heats atmosphere above spot

Cooler gases in atmosphere sink

Hot gases rise from storm

ENERGY TRANSFER

Gases are spun together by planet's spin

Spot is constantly changing, with material entering and exiting

HOW STRONG ARE THE WINDS ON JUPITER?

Surface winds on Jupiter can blow at over 370 mph (600 kph). It is thought that these winds are driven by convection deep inside Jupiter's hot interior.

GREAT RED SPOT

Eddies crash together at base of storm, transferring energy

Eddies join together, feeding storm with energy

Without a solid surface, there is less friction to slow storm

Jupiter's weather

No other planet has weather quite like Jupiter's. Its atmosphere churns with colossal storms and is riddled with lightning, both more powerful than anything experienced on Earth.

Cloud layers

Jupiter's visible surface is striped with orange, red, brown, and white clouds. Cyclones press together at Jupiter's poles, and eddies and whirlpools swirl around the planet, some spinning against Jupiter's rotation and persisting for centuries. The upper layers of Jupiter's clouds are laced with white ammonia ice and are organized into stripes called zones, which sit parallel to the planet's equator. Where these clouds are absent, deeper layers of the Jovian atmosphere are exposed, resulting in darker bands called belts.

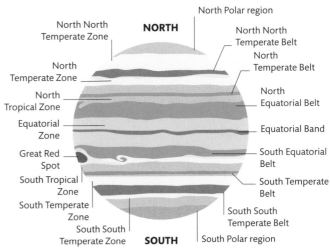

North Polar region
North North Temperate Zone
NORTH
North North Temperate Belt
North Temperate Belt
North Temperate Zone
North Equatorial Belt
North Tropical Zone
Equatorial Zone
Equatorial Band
Great Red Spot
South Equatorial Belt
South Tropical Zone
South Temperate Belt
South Temperate Zone
South South Temperate Belt
South South Temperate Zone
SOUTH
South Polar region

Zones and belts
The weather on Jupiter is driven by convection, with hot gas rising within the white zones and cooler gas falling in the darker belts.

LIGHTNING IN JUPITER'S ATMOSPHERE STRIKES UP TO FOUR TIMES PER SECOND

JUPITER'S LIGHTNING

Lightning was first spotted on Jupiter in 1979 by the Voyager 1 spacecraft. These flashes, which typically appear near Jupiter's poles, are more powerful than lightning on Earth. Water vapor rises through Jupiter's interior and forms droplets in the atmosphere. Higher up, where it is colder, the droplets freeze. Electric charge builds up as the droplets collide in the cloud layers, then discharges as lightning.

Lightning discharges inside cloud layer

Ice particles and water droplets separate out

LIQUID HYDROGEN LAYER

METALLIC HYDROGEN LAYER

Water vapor rises from interior

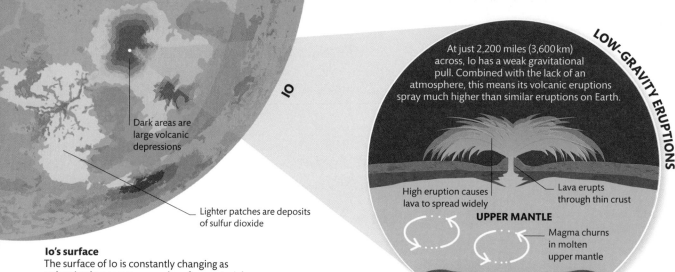

Dark areas are large volcanic depressions

Lighter patches are deposits of sulfur dioxide

Io's surface
The surface of Io is constantly changing as volcanic plumes spew up subsurface material, creating lava lakes, mountains, and volcanoes that can stretch up to 155 miles (250 km) across.

LOW-GRAVITY ERUPTIONS

At just 2,200 miles (3,600 km) across, Io has a weak gravitational pull. Combined with the lack of an atmosphere, this means its volcanic eruptions spray much higher than similar eruptions on Earth.

High eruption causes lava to spread widely

Lava erupts through thin crust

UPPER MANTLE

Magma churns in molten upper mantle

Magma rises through solid lower mantle

LOWER MANTLE

Io and Europa

Jupiter has 79 moons, two of which are among the most exciting yet contrasting moons in the Solar System. Both Io and Europa are shaped by Jupiter's immense gravitational pull.

Galilean satellites

Io and Europa are two of Jupiter's four largest moons, called the Galilean satellites. At a distance of only 260,000 miles (420,000 km) from Jupiter, Io's close orbit takes just 1.5 days to complete. As it does so, Io experiences huge tides that turn it into the most volcanically active place in the Solar System. Meanwhile, Europa is farther away, taking 3.5 days to orbit Jupiter. It has less tidal heating but enough to make an ocean of water under the hard, icy crust.

Hottest spots often last only a few days

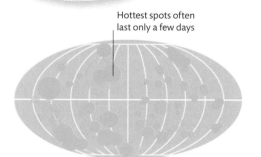

ERUPTIONS ON IO

Volcanic map
When mapped, Io's volcanic hot spots appear to be located at random, but they are more widely spaced at the moon's equator. Tectonic activity may be driving these areas apart.

TIDAL HEATING

Because Io travels on an elliptical orbit, its distance from Jupiter varies. As a result, the tidal forces due to Jupiter's gravity change, too, constantly stretching and squeezing Io. This input of energy heats up the interior. Tidal heating affects all the Galilean satellites.

Weaker pull as moon moves away

Jupiter's pull is weaker on far side

IO

JUPITER

Tides pulled toward Jupiter

Strong pull close to planet

HOW MUCH DOES JUPITER STRETCH IO?

Jupiter's gravity and Io's elliptical orbit both cause the moon's surface to bulge. Its solid surface stretches by up to 330 ft (100 m) every 1.5 days.

Activity on Europa
Eruptions of liquid water and water vapor have been seen on Europa's surface. It is thought that water in the subsurface ocean, heated by tidal forces due to Jupiter, rises to the crust and bursts through the surface.

EUROPA HAS THE SMOOTHEST SURFACE OF ANY SOLID SOLAR SYSTEM BODY

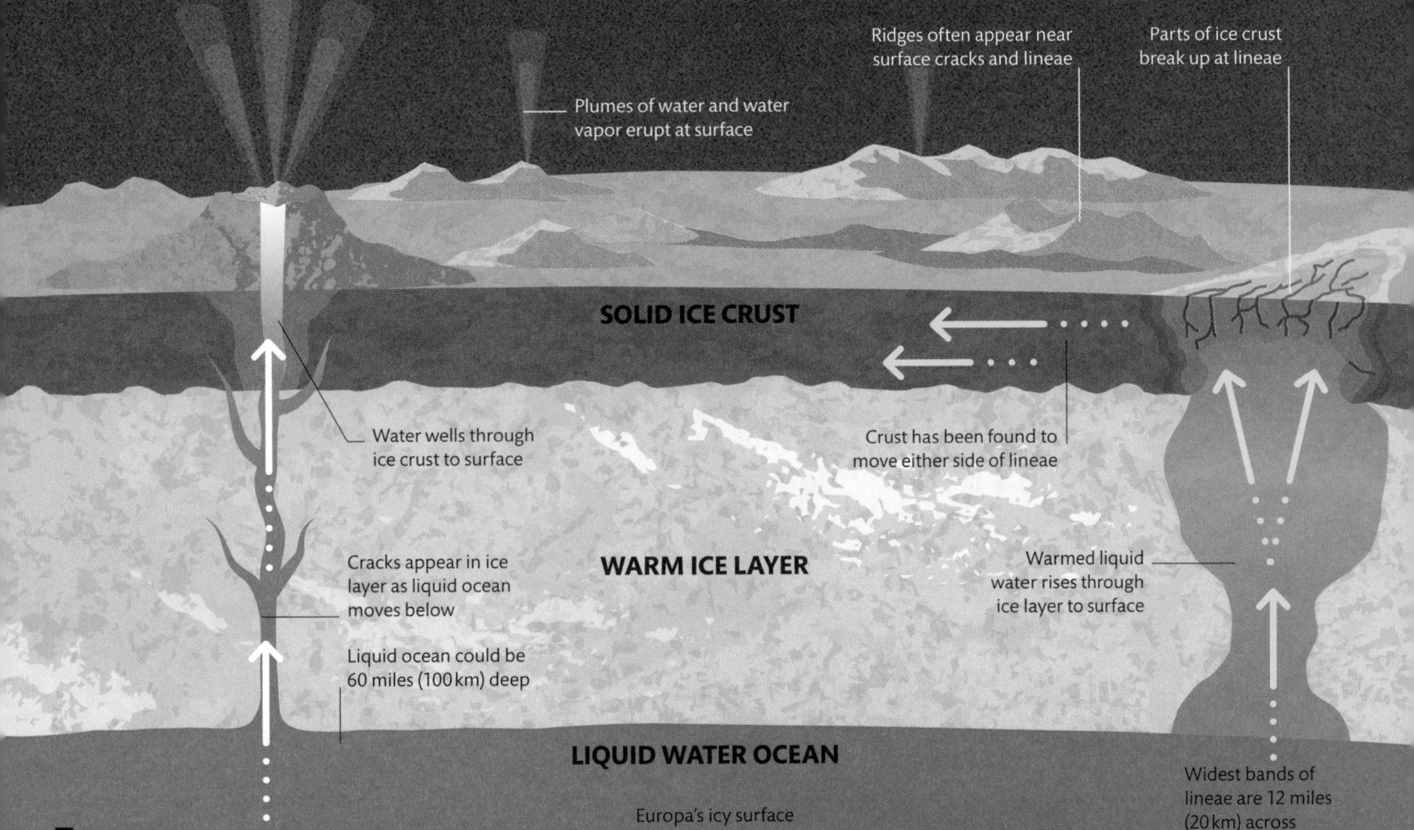

Ridges often appear near surface cracks and lineae

Parts of ice crust break up at lineae

Plumes of water and water vapor erupt at surface

SOLID ICE CRUST

Water wells through ice crust to surface

Crust has been found to move either side of lineae

Cracks appear in ice layer as liquid ocean moves below

WARM ICE LAYER

Warmed liquid water rises through ice layer to surface

Liquid ocean could be 60 miles (100 km) deep

LIQUID WATER OCEAN

Widest bands of lineae are 12 miles (20 km) across

Europa's icy surface is highly reflective

Europa

Europa's solid ice crust is streaked with lines, and there is much debate about how thick the ice is. Beneath the ice is an ocean that contains more liquid water than all of Earth's oceans, seas, lakes, and rivers combined, leading some scientists to believe it is a potential site to explore for signs of life. Under the ocean is a layer of rock on top of a metallic core.

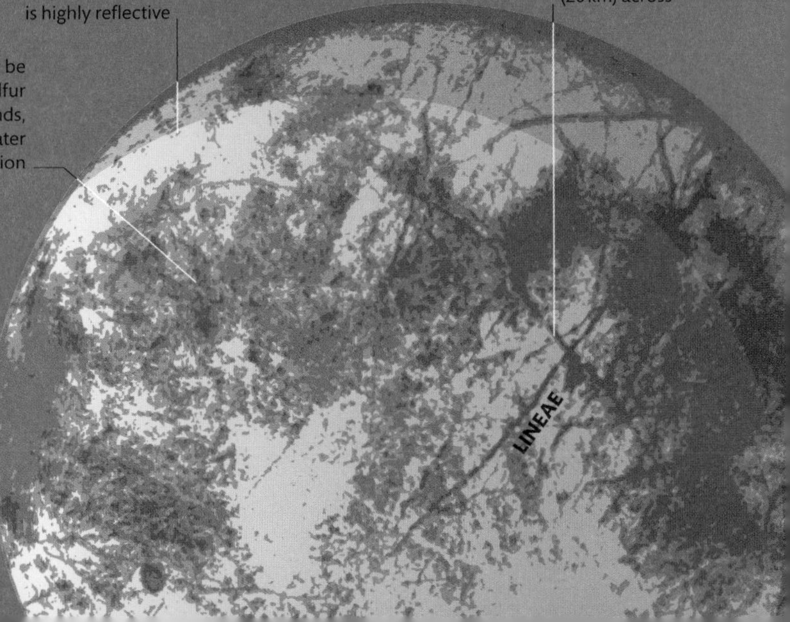

Dark spots may be salts and sulfur compounds, affected by water ice and radiation

Europa's surface
Dark streaks on Europa's surface, called lineae, are thought to be caused by the movement of water underneath. Similar features are seen close to Earth's ice caps.

EUROPA

LINEAE

Ganymede and Callisto

The outer two Galilean satellites, Ganymede and Callisto, are larger and less active than Europa and Io. They are also scarred from billions of years' worth of high-energy impacts.

IF GANYMEDE IS SO BIG, WHY ISN'T IT A PLANET?

Although Ganymede is round and bigger than Mercury, it is not classified as a planet. All planets must orbit the Sun, but Ganymede orbits Jupiter.

Ganymede

At 3,300 miles (5,300 km) wide, Ganymede is the biggest moon in the Solar System and larger than Mercury (although Ganymede is not as heavy). It has a thin atmosphere largely composed of oxygen and is also the only satellite known to have its own magnetic field, indicating that it has an iron core and distinct internal layers. Ganymede orbits Jupiter in a week and always shows the same side to its host. The moon's surface alternates between dark, cratered regions and light patches with ridges that may stem from tectonic activity.

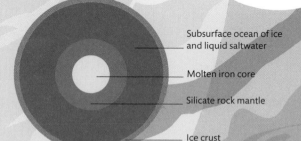

Subsurface ocean of ice and liquid saltwater

Molten iron core

Silicate rock mantle

Ice crust

JUPITER

Inside Ganymede
Ganymede has a liquid iron core with a temperature in excess of 2,700°F (1,500°C). This warms a layer of silicate rock and a vast subsurface ocean containing more water than on Earth. The surface is formed of a hard ice shell.

Callisto

Callisto—only a little smaller than Mercury—is home to the Solar System's most heavily cratered surface. The impacts are very old and distinct, suggesting the moon's surface has not been altered by volcanic or tectonic activity for over 4 billion years. Callisto is also the only Galilean satellite not to undergo significant tidal heating. At almost 1.2 million miles (1.9 million km) from Jupiter, Callisto is the most distant major moon and is also less affected by Jupiter's powerful magnetosphere.

VALHALLA

Prominent impact craters located inside basin

Concentric rings surround bright center

CALLISTO IS THE **MOST HEAVILY CRATERED** OBJECT IN THE **SOLAR SYSTEM**

Multiring impact crater
Callisto has the largest multiring impact basin in the Solar System, called Valhalla. It stretches 2,400 miles (3,800 km) across.

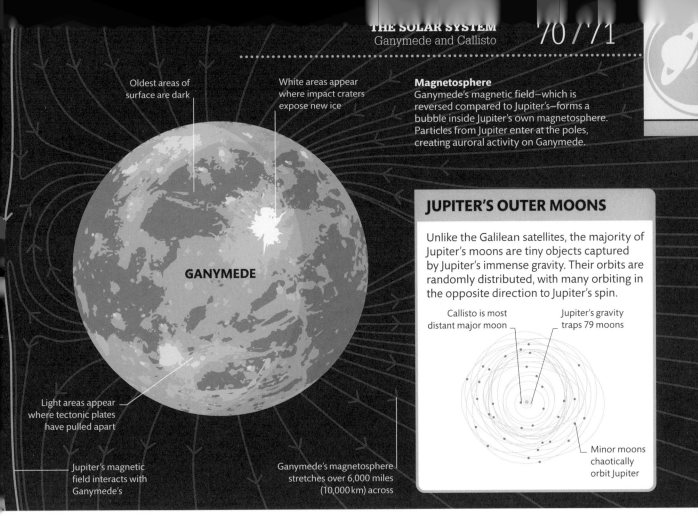

Oldest areas of surface are dark

White areas appear where impact craters expose new ice

Magnetosphere
Ganymede's magnetic field—which is reversed compared to Jupiter's—forms a bubble inside Jupiter's own magnetosphere. Particles from Jupiter enter at the poles, creating auroral activity on Ganymede.

GANYMEDE

Light areas appear where tectonic plates have pulled apart

Jupiter's magnetic field interacts with Ganymede's

Ganymede's magnetosphere stretches over 6,000 miles (10,000 km) across

JUPITER'S OUTER MOONS

Unlike the Galilean satellites, the majority of Jupiter's moons are tiny objects captured by Jupiter's immense gravity. Their orbits are randomly distributed, with many orbiting in the opposite direction to Jupiter's spin.

Callisto is most distant major moon

Jupiter's gravity traps 79 moons

Minor moons chaotically orbit Jupiter

Typical crater formation

Many craters seen across the Solar System are created by large impacts, the force of which melts both the impactor and the impact site. After the initial shock effect, molten material rises and solidifies in the middle of the crater and debris is often ejected and scattered around the edge of the crater. Chains of small craters are the result of impacts from comets torn into pieces by the moon's tidal forces.

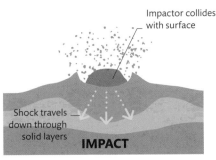

Impactor collides with surface

Shock travels down through solid layers

IMPACT

Debris scatters around edge of basin

Fractured rock settles in basin

CRATER FORMS

Valhalla formation

This distinctive structure of concentric rings formed when an impact completely punctured the outer shell of Callisto's surface, exposing softer material below that may have been an ocean. This deeper material flowed toward the center of the crater, filling up the space carved out by the impact. As the softer material moved, surface material around the edge of the crater collapsed, forming the rings.

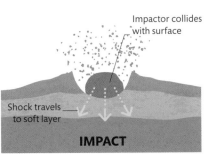

Impactor collides with surface

Shock travels to soft layer

IMPACT

Edge collapses as material below moves

Soft layer beneath surface exposed

RINGS FORM

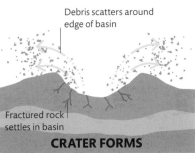

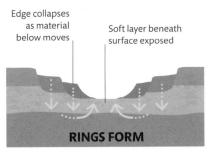

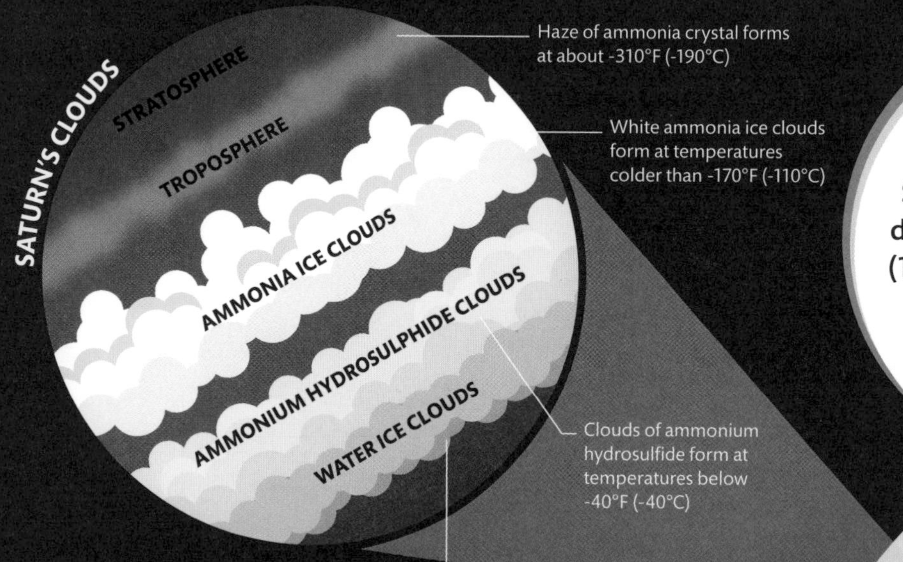

SATURN'S CLOUDS

STRATOSPHERE

TROPOSPHERE

Haze of ammonia crystal forms
at about -310°F (-190°C)

White ammonia ice clouds
form at temperatures
colder than -170°F (-110°C)

AMMONIA ICE CLOUDS

AMMONIUM HYDROSULPHIDE CLOUDS

WATER ICE CLOUDS

Clouds of ammonium
hydrosulfide form at
temperatures below
-40°F (-40°C)

Cloud layers
The atmosphere is formed of hydrogen;
helium; and traces of ammonia, methane,
and water vapor. Cold temperatures create
layers of ice clouds as the gases freeze.

Water ice and vapor
clouds form at 32°F
(0°C) or colder

**HOW FAR
IS SATURN FROM
THE SUN?**

Saturn orbits at an average
distance of 890 million miles
(1.4 billion km) from the Sun.
It takes 80 minutes for
sunlight to reach Saturn,
10 times longer
than for Earth.

Saturn

Saturn is the sixth planet from the Sun
and the second largest planet in the Solar
System. It is best known for its famous
ring system.

The ringed planet

Saturn is a gas giant, formed of mostly hydrogen
and helium, meaning that—unlike Earth or any of
the other rocky planets—it has no real surface. With
a radius of 36,000 miles (58,000km), Saturn is nine
times wider than Earth. While it is renowned for its
rings, made almost entirely of ice, Saturn is not
the only planet with rings. In fact, all four
giant planets have them, but only
Saturn's are clearly visible.

Inside Saturn

Scientists think that deep inside Saturn,
under miles (km) of gaseous atmosphere,
is a layer of liquid molecular hydrogen.
Below this, the hydrogen is under such
pressure that the molecules break down
into atoms and turn into a conductive liquid
called metallic hydrogen. At the center of
the planet is a dense core, with a
temperature up to 18,000°F (10,000°C),
which might be solid or liquid.

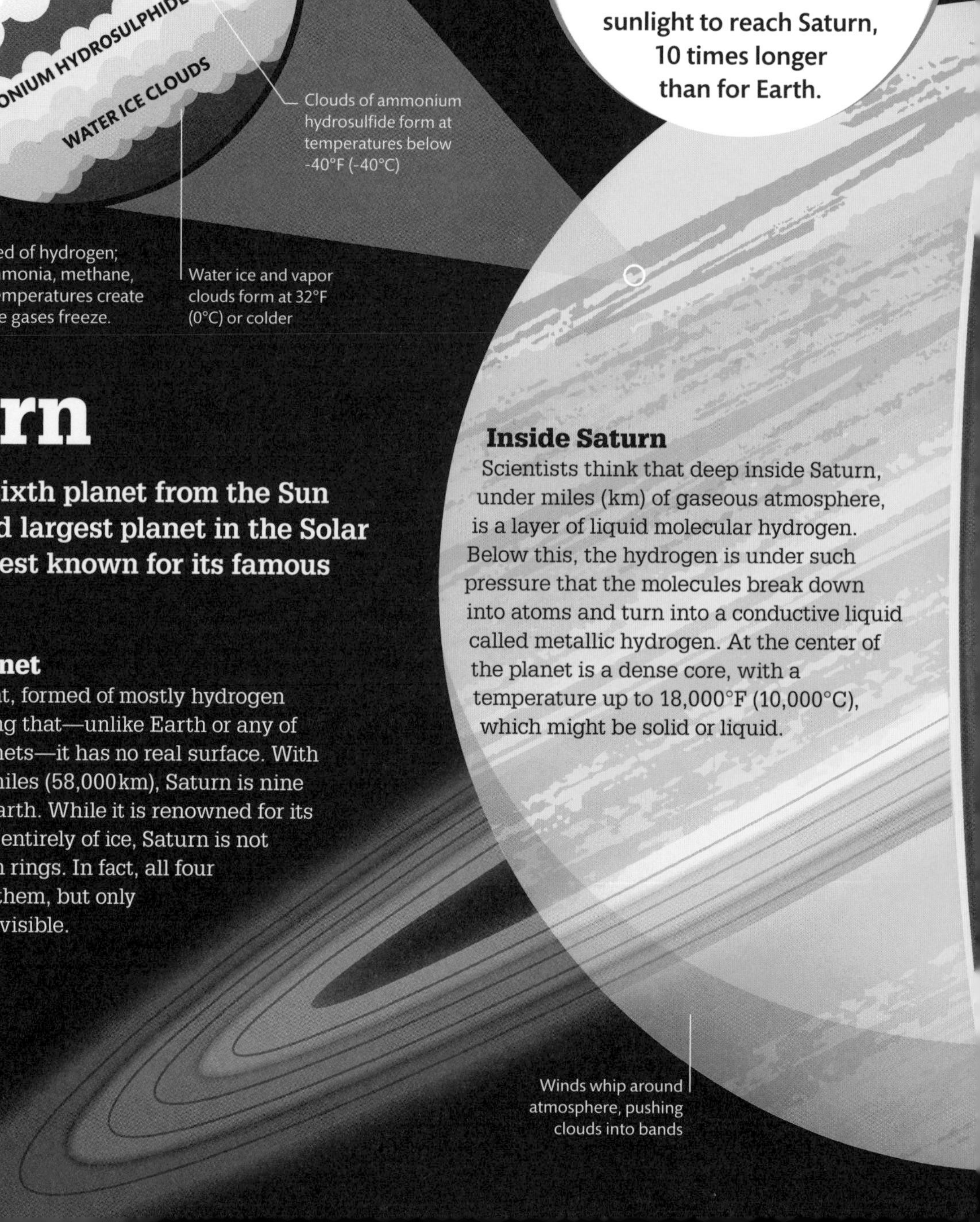

Winds whip around
atmosphere, pushing
clouds into bands

Hexagonal vortex
Near Saturn's north pole is a hexagonal cloud pattern, or vortex, with each side around 9,000 miles (14,500 km) long. It is thought to be caused by complex turbulence in the atmosphere.

Swirling clouds of vortex

NORTH POLE TURBULENCE

Ring system extends up to 175,000 miles (282,000 km) from planet

Saturn's rings might be icy fragments from a moon that was destroyed in a collision

Liquid starts becoming metallic

Dense, hot core may contain rock and metal

ROCKY CORE

METALLIC HYDROGEN

MOLECULAR HYDROGEN

Liquid metallic layer is source of Saturn's magnetic field

Liquid layer consisting of hydrogen and helium under pressure

SATURN'S DENSITY IS SO LOW THAT THE PLANET WOULD FLOAT IN WATER

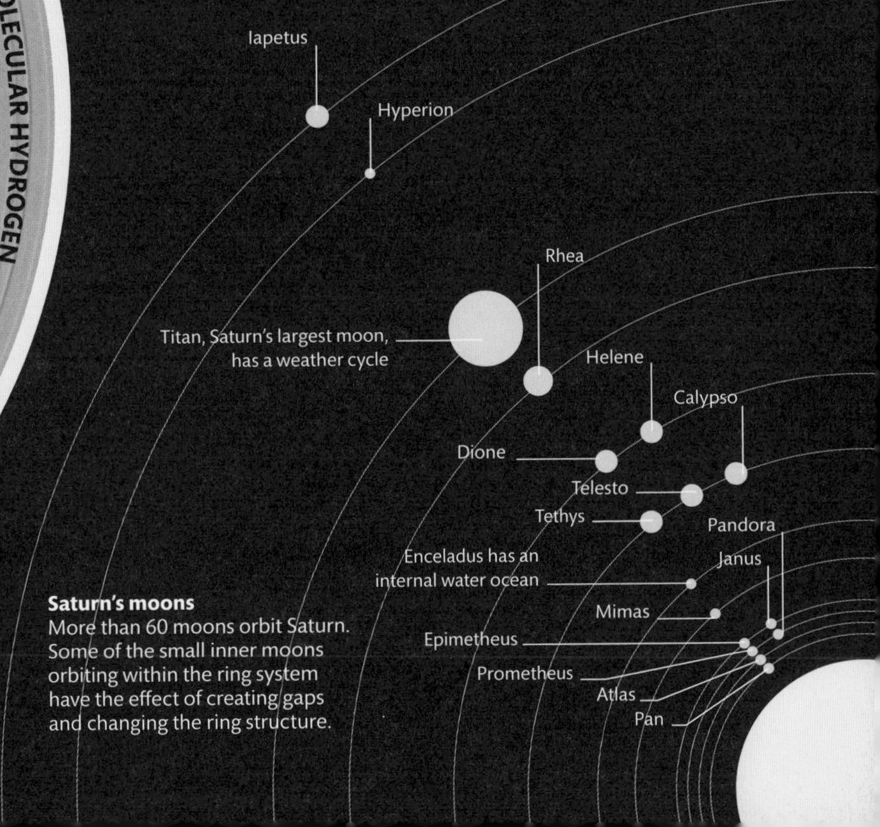

Troposphere layer of atmosphere is Saturn's visible surface

Iapetus

Hyperion

Rhea

Helene

Calypso

Titan, Saturn's largest moon, has a weather cycle

Dione

Telesto

Tethys

Pandora

Janus

Enceladus has an internal water ocean

Mimas

Epimetheus

Prometheus

Atlas

Pan

Internal layers
Saturn's internal layers are formed of around 75 percent hydrogen and 25 percent helium. The layers change gradually as pressure builds closer to the core.

Saturn's moons
More than 60 moons orbit Saturn. Some of the small inner moons orbiting within the ring system have the effect of creating gaps and changing the ring structure.

The inner rings

Saturn's rings are identified by letters that were allocated in the order in which they were discovered. The two brightest are the A and B rings, separated by the Cassini Division. Extending inward from the B ring are the paler C and D rings, which contain smaller ice particles.

SATURN'S RINGS MIGHT HAVE FORMED ONLY 10–100 MILLION YEARS AGO—AFTER LIFE BEGAN ON EARTH

Rings have complex structure of gaps and ringlets

E ring's tiny particles make it almost invisible

E RING

G ring is made of very fine particles

F ring is most active, changing every few hours

D RING

Maxwell Gap has a narrow ringlet inside

Columbo Gap is found in inner C ring

C RING

B RING

A RING

F RING

G RING

B ring is largest, brightest, and most massive

Encke Division is a 200-mile (325-km) wide gap inside A ring

16 FT (5 M) DEEP

16–33 FT (5–10 M) DEEP

33–98 FT (10–30 M) DEEP

Innermost ring is extremely faint

Faint, dark C ring is 11,000 miles (17,500 km) wide

Pull of moon Mimas causes Cassini Division

F ring is outermost of large, bright rings

The outer rings

Beyond the distinct D to G rings is a series of extremely wide, faint outer rings that stretch out to the orbit of Saturn's moon Phoebe. The E ring is faintly visible, but the outermost ring, called the Phoebe ring because it stretches out to the moon Phoebe, is made up of particles so small that the ring is almost invisible.

SATURN

MIMAS

ENCELADUS

TETHYS

DIANA

RHEA

Faint outer rings

TO PHOEBE

DISTANCE TO OUTERMOST RING

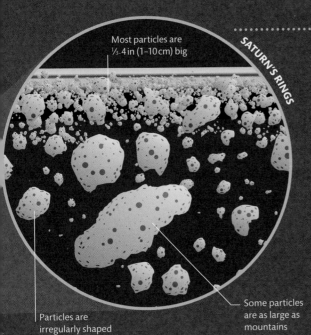

Most particles are ⅓–4 in (1–10 cm) big

SATURN'S RINGS

Particles are irregularly shaped

Some particles are as large as mountains

Ring materials

Saturn's rings are almost entirely made up of water ice, with some bits of dust and rock from passing comets, asteroids, and the impacts of meteorites on Saturn's moons. The chunks of ice in the rings range in size from dust particles to miles wide. The densest areas are inside the A and B rings, which were the first to be discovered as the high density of chunks they contain makes them more visible.

Ice particles

Inside, the particles are over 99.9 percent water ice, with trace components of rocky materials. These materials include silicates and tholins, which are organic compounds created by cosmic rays interacting with hydrocarbons like methane.

Saturn's rings

While the bright ring system around Saturn may look solid, the rings are in fact made of countless chunks of almost pure water ice, orbiting the gas giant in a series of distinct rings.

WHAT COLOR ARE THE RINGS?

Saturn's rings look whitish because they are almost entirely water ice. However, NASA's Cassini spacecraft distinguished pale shades of pink, gray, and brown due to impurities.

The ring system

Icy chunks forming the iconic rings of Saturn may be debris from a moon that broke up or may even be left over from the giant planet's formation. Over time, these chunks were covered in layers of dust and started to orbit the planet. Saturn's rings have a typical thickness of 33–66 ft (10–20 m) but can reach a thickness of up to 0.6 miles (1 km). The inner rings stretch out 109,000 miles (175,000 km) from Saturn and are separated by gaps caused by the gravitational pull of Saturn's moons. The largest gap, the Cassini Division, is 2,900 miles (4,700 km) wide.

HOW THE RINGS FORMED

Exactly how Saturn's rings formed remains uncertain. A popular idea is that one of Saturn's moons moved in toward Saturn and broke up when it crossed the Roche limit, the point where the planet's tidal forces could tear it apart. In one theory, the ring pieces broke off the icy mantle of a large moon and then the rocky core of the moon spiraled into Saturn.

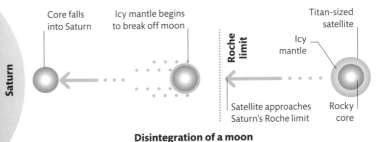

Core falls into Saturn

Icy mantle begins to break off moon

Saturn

Roche limit

Titan-sized satellite

Icy mantle

Satellite approaches Saturn's Roche limit

Rocky core

Disintegration of a moon

Inside Titan

Information gathered by NASA's Cassini spacecraft indicates Titan's internal structure is made up of five layers. In the center is a core of silicate rock around 2,500 miles (4,000 km) in diameter. This is surrounded by a shell of ice-VI, a type of water ice that forms under high pressures. Above this is a layer of salty liquid water, followed by a layer of water ice. The outermost layer, Titan's surface, is made up of hydrocarbons (organic compounds of hydrogen and carbon) that have accumulated in the form of sands or liquids. A dense, high-pressure atmosphere extends 370 miles (600 km) above the surface into space.

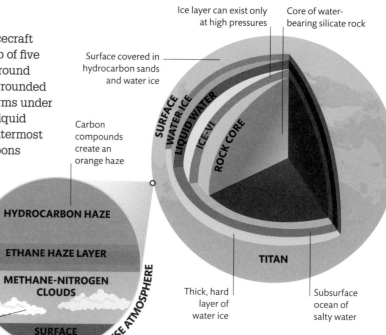

Ice layer can exist only at high pressures

Core of water-bearing silicate rock

Surface covered in hydrocarbon sands and water ice

Carbon compounds create an orange haze

SURFACE
WATER-ICE
LIQUID WATER
ICE-VI
ROCK CORE

TITAN

HYDROCARBON HAZE

ETHANE HAZE LAYER

METHANE-NITROGEN CLOUDS

SURFACE

DENSE ATMOSPHERE

Ethane haze formed by solar radiation

Methane and nitrogen molecules form low clouds

Thick, hard layer of water ice

Subsurface ocean of salty water

Atmospheric elements
Titan's atmosphere is composed of around 95 percent nitrogen and 5 percent methane, with small amounts of organic compounds rich in hydrogen and carbon.

Titan's weather

Titan's surface is one of the most Earth-like places in the Solar System, but it is much colder. Temperatures are around -290°F (-180°C), as the surface receives about 1 percent of the light that reaches Earth. Titan's weather cycle sees hydrocarbons, such as methane and ethane, cooled to the point of liquidity to form rain, rivers, and seas. The cycle starts with the accumulation of methane and nitrogen in the thick atmosphere.

Organic compounds form in atmosphere, creating clouds

PRECIPITATION

Compounds condense into raindrops and fall to ground

Methane enters atmosphere through volcanoes or cracks in surface

HOW MUCH BIGGER IS TITAN THAN EARTH'S MOON?

Titan's diameter is 50 percent larger than Earth's Moon, at 3,200 miles (5,150 km). Titan is also 80 percent heavier thanks to its dense silicate core.

1 Organic compounds form
Methane from below the surface leaks out to the atmosphere. At high altitude, methane and nitrogen molecules are split apart by ultraviolet light from the Sun. The atoms then recombine to form organic compounds containing hydrogen and carbon.

2 Rain brings down compounds
Some of the organic compounds accumulate in clouds, then fall to the ground as rain. The low gravity and dense atmosphere of Titan causes rain to fall at about 4 mph (6 kph), about six times slower than on Earth.

Titan

Saturn's biggest moon, and the second largest moon in the Solar System after Ganymede, Titan has clouds and rain and is covered in lakes. Titan is the only body in the Solar System with a cycle similar to Earth's water cycle. However, in Titan's case, it rains methane.

TITAN IS 3,200 MILES (5,150 KM) WIDE, BIGGER THAN THE PLANET MERCURY

IDENTIFYING TITAN'S LAKES

NASA's Cassini spacecraft used radar to map surface features and bodies of liquid methane and ethane on Titan. The way infrared radiation was absorbed or reflected also helped identify liquid.

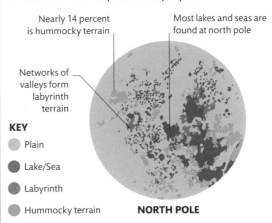

Nearly 14 percent is hummocky terrain

Most lakes and seas are found at north pole

Networks of valleys form labyrinth terrain

KEY

- Plain
- Lake/Sea
- Labyrinth
- Hummocky terrain

NORTH POLE

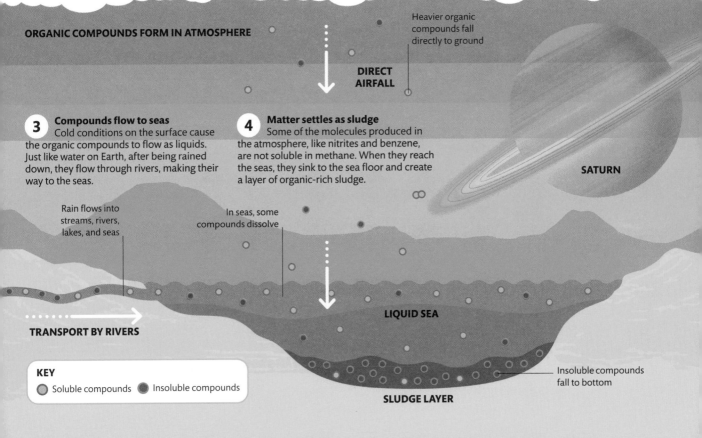

ORGANIC COMPOUNDS FORM IN ATMOSPHERE

Heavier organic compounds fall directly to ground

DIRECT AIRFALL

3 **Compounds flow to seas**
Cold conditions on the surface cause the organic compounds to flow as liquids. Just like water on Earth, after being rained down, they flow through rivers, making their way to the seas.

4 **Matter settles as sludge**
Some of the molecules produced in the atmosphere, like nitrites and benzene, are not soluble in methane. When they reach the seas, they sink to the sea floor and create a layer of organic-rich sludge.

SATURN

Rain flows into streams, rivers, lakes, and seas

In seas, some compounds dissolve

LIQUID SEA

TRANSPORT BY RIVERS

Insoluble compounds fall to bottom

KEY
- Soluble compounds
- Insoluble compounds

SLUDGE LAYER

Ice giants

There are two giant ice planets—Uranus and Neptune—located in the outer Solar System. These large planets are made mostly of water, ammonia, and methane.

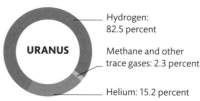

URANUS

Hydrogen: 82.5 percent

Methane and other trace gases: 2.3 percent

Helium: 15.2 percent

Uranus

Uranus, the seventh planet from the Sun, orbits slowly at a distance of around 1.8 billion miles (2.9 billion km), but it rotates quickly, taking around 17 hours to complete one rotation around its axis. At 32,000 miles (51,000 km) across, Uranus is about four times the width of Earth and has 27 moons and 13 barely visible rings. Unlike most other planets, Uranus rotates east to west, possibly the result of a collision with an Earth-sized object.

Inside Uranus
Beneath a deep atmosphere, Uranus gets most of its mass from a liquid mantle of water, ammonia, and methane—called "ices" because they are normally frozen in the outer Solar System. This surrounds a small, rocky core. Although Uranus's atmosphere is cold, its core may reach almost 9,000°F (5,000°C).

Upper atmosphere forms Uranus's visible surface

Atmospheric composition
Uranus's atmosphere is composed mostly of hydrogen and helium, with a small amount of methane and traces of water and ammonia. Neptune's atmosphere has an almost identical composition.

RINGS

Two outer rings are broad and faint

Rings composed of dark particles made of ice and rock

Inner rings consist of nine narrow rings and two dusty rings

UPPER ATMOSPHERE

LOWER ATMOSPHERE

Mixture of water, ammonia, and methane ices form mantle

MANTLE

Strong winds circulate in lower atmosphere

Mantle is a dense, hot liquid because of high temperatures

Uranus's core is mainly rock

CORE

WHY ARE THE ICE GIANTS BLUE?

Methane in the atmospheres of both planets absorbs red sunlight, so the reflected light looks blue. Neptune's darker color suggests there is another unknown chemical in its atmosphere.

Light streaks of ammonia cloud

Neptune

Neptune is the outermost planet in the Solar System, at a distance of about 2.8 billion miles (4.5 billion km) from the Sun. While it also appears blue, it is a darker shade than Uranus, and its clouds and a dark spot are signs of an active atmosphere. The movement of clouds on the visible surface have shown that Neptune has the strongest winds in the Solar System. Neptune is slightly smaller than Uranus, with 14 known moons and at least five rings.

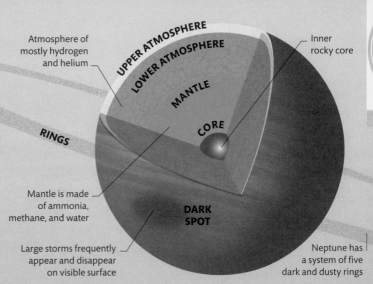

Atmosphere of mostly hydrogen and helium

UPPER ATMOSPHERE

LOWER ATMOSPHERE

MANTLE

Inner rocky core

CORE

RINGS

Mantle is made of ammonia, methane, and water

DARK SPOT

Large storms frequently appear and disappear on visible surface

Neptune has a system of five dark and dusty rings

Inside Neptune

Like Uranus, Neptune's interior is made up of a core of rock and ice, followed by a mantle of water, ammonia, and methane ice. There might also be an ocean of super-hot water under Neptune's clouds.

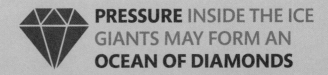

PRESSURE INSIDE THE ICE GIANTS MAY FORM AN **OCEAN OF DIAMONDS**

Supersonic winds

Neptune's strong winds whirl around the planet at speeds 1.5 times the speed of sound. Gravitational studies show these high-speed winds are contained in the upper atmosphere.

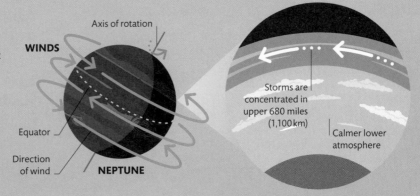

Axis of rotation

WINDS

Equator

Direction of wind

NEPTUNE

Storms are concentrated in upper 680 miles (1,100 km)

Calmer lower atmosphere

Sun's ultraviolet light interacting with atmosphere gives a hazy appearance

URANUS'S UNUSUAL SEASONS

Uranus's equator is nearly at a right angle to the its orbital plane, with a tilt of almost 98°, possibly caused by a collision with a large object soon after the planet's formation. As a result, Uranus has the most extreme seasons of any planet in the Solar System. A quarter of Uranus's orbit, 21 years, is spent with one pole facing the Sun and the other in darkness.

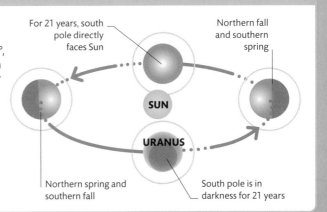

For 21 years, south pole directly faces Sun

Northern fall and southern spring

SUN

URANUS

Northern spring and southern fall

South pole is in darkness for 21 years

Pluto

Originally classified as a planet, Pluto was reclassified as a dwarf planet when similar worlds were discovered in the outer Solar System. This cold dwarf planet has a complex terrain with mountains and ice plains.

Surface features

Pluto is one of the larger dwarf planets, but it has a diameter of only 1,400 miles (2,300 km)—about two-thirds the size of Earth's moon. It orbits the Sun at an average distance of 3.7 billion miles (5.9 billion km), hence the cold surface temperatures. Pluto's surface is covered in mountains, valleys, and ice plains, the most distinctive of which is the ice plain Sputnik Planitia. Stretching 600 miles (1,000 km) across, this plain formed when a Kuiper Belt object collided with Pluto.

Kuiper Belt object 30–60 miles (50-100 km) across collided with Pluto

Large area of icy crust was removed

Thin, weak layer of crust was left behind

Ocean beneath surface pushed against weak layer, extending scarring

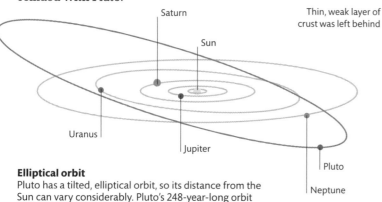

Saturn

Sun

Uranus

Jupiter

Pluto

Neptune

Elliptical orbit
Pluto has a tilted, elliptical orbit, so its distance from the Sun can vary considerably. Pluto's 248-year-long orbit takes it as far as 4.6 billion miles (7.4 billion km) from the Sun and as close as 2.7 billion miles (4.4 billion km).

PLUTO'S MOONS

Pluto is orbited by five moons, formed by a collision between Pluto and a similarly sized body. The largest moon, Charon, is around half Pluto's size and so similar that they are sometimes considered to be a double-planet system.

HYDRA

KERBEROS

NIX

STYX

CHARON

PLUTO

Sputnik Planitia
A large object colliding with Pluto and exposing the crust may have created its most prominent feature. Ice slush from a subsurface ocean and frozen nitrogen then formed plains, troughs, and hills.

PLUTO'S ORBIT BRINGS IT CLOSER TO THE SUN THAN NEPTUNE

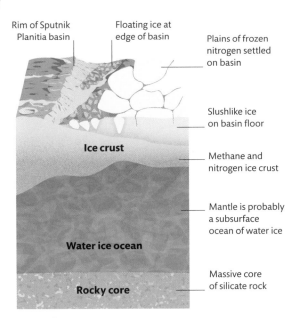

Rim of Sputnik Planitia basin

Floating ice at edge of basin

Plains of frozen nitrogen settled on basin

Slushlike ice on basin floor

Ice crust

Methane and nitrogen ice crust

Mantle is probably a subsurface ocean of water ice

Water ice ocean

Massive core of silicate rock

Rocky core

Internal structure
Pluto's crust is formed of an ice sheet at least 2.5 miles (4 km) thick. This sheet covers a possible liquid water ocean and a large rocky core that forms 60 percent of Pluto's mass.

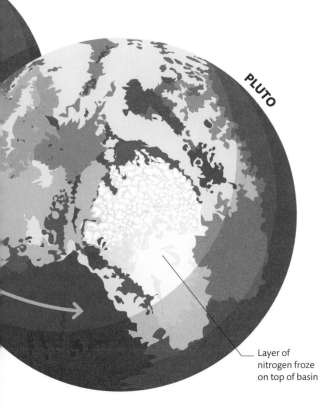

PLUTO

Layer of nitrogen froze on top of basin

HOW OLD IS PLUTO?

Like most things in the Kuiper Belt, Pluto formed in the very early Solar System, about 4.5 billion years ago. The collision that formed Sputnik Planitia occurred 4 billion years ago.

Pluto's volcanoes

To the south of Sputnik Planitia, there are two huge, strange-looking mountains. The larger one, Piccard Mons, is 4 miles (7 km) high and 140 miles (225 km) wide. It is thought they may be cryovolcanoes. Instead of an eruption of molten rock, cryovolcanoes send liquids or vapors of chemicals such as water, ammonia, and methane into the atmosphere. They occur in places where the surrounding temperature is extremely cold.

Cloud of vapor and liquid erupts through surface

Ejected material begins to refreeze

Melted materials rise up through icy surface

Frozen material builds on surface, forming a mountain

Frozen chemicals in liquid ocean melt

ICE CRUST

LIQUID OCEAN

ROCKY CORE

How cryovolcanoes work
Frozen chemicals beneath the surface are heated by radioactive decay or tidal forces. The chemicals melt and erupt to the surface, where they rapidly refreeze.

Rocky core may heat up

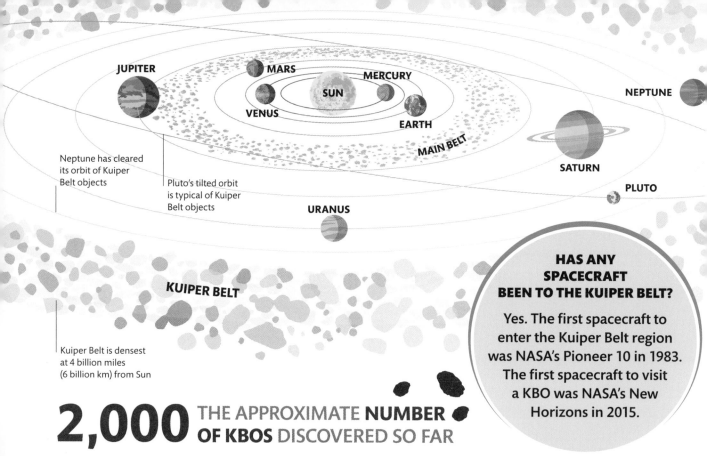

Neptune has cleared
its orbit of Kuiper
Belt objects

Pluto's tilted orbit
is typical of Kuiper
Belt objects

KUIPER BELT

Kuiper Belt is densest
at 4 billion miles
(6 billion km) from Sun

2,000 THE APPROXIMATE **NUMBER** OF **KBOS** DISCOVERED SO FAR

HAS ANY SPACECRAFT BEEN TO THE KUIPER BELT?

Yes. The first spacecraft to enter the Kuiper Belt region was NASA's Pioneer 10 in 1983. The first spacecraft to visit a KBO was NASA's New Horizons in 2015.

The Kuiper Belt

In the outer part of the Solar System, extending beyond the orbit of Neptune, is a donut-shaped ring of icy objects called the Kuiper Belt.

How the Kuiper Belt formed

The planets in the Solar System formed when gas, dust, and rocks pulled together under gravity. Beyond the planets, a disk of debris was left. Over time, the planets Saturn, Uranus, and Neptune migrated outward. The giant planet Neptune, orbiting close to the disk of debris, disturbed the orbits of objects inside it. Neptune's gravity scattered many of them farther from the Sun, into the Oort Cloud (pp.84–85) or out of the Solar System completely. In the end, only a small fraction of the original number of objects was left. Even so, many millions of small, icy bodies are believed to remain in the Kuiper Belt region.

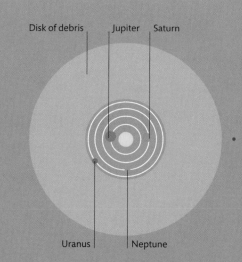

1 **Compact ring of debris**
Objects in the Kuiper Belt, along with Neptune and Uranus, are thought to have formed closer to the Sun than they are now. The objects may have come from a disk of protoplanetary debris near the planets.

Kuiper Belt objects (KBOs)

There are potentially millions of icy objects floating around in the Kuiper Belt. They are generally white, but their color can change to red as a result of solar radiation.

Frozen Kuiper Belt objects have a temperature around -360°F (-220°C)

The icy belt

Extending from the orbit of Neptune, at about 2.8 billion miles (4.5 billion km) from the Sun, to 5 billion miles (8 billion km), the Kuiper Belt is similar to the Main Belt (see pp.60–61) but much bigger. Being so far from the Sun, it is a cold and dark place. It is home to hundreds of thousands of icy objects more than 60 miles (100 km) across, made up mostly of frozen ammonia, water, and methane. Some have moons, and they include larger objects classed as dwarf planets. The Kuiper Belt is also the area where some comets originate (see pp.84–85).

DWARF PLANETS

Four of the largest objects beyond Neptune are classed as dwarf planets. Dwarf planets orbit the Sun and have become rounded under the force of their own gravity, but they are not large enough to clear other objects out of their orbit.

Pluto
At 1,500 miles (2,400 km) across, Pluto is the largest dwarf planet.

Eris
Eris is very slightly smaller than Pluto, but it is more massive.

Makemake
Makemake is about two-thirds the size of Pluto and has a small moon.

Haumea
Egg-shaped Haumea has two moons and a ring system around it.

Ceres
Ceres, in the Main Belt, is the only dwarf planet not orbiting beyond Neptune.

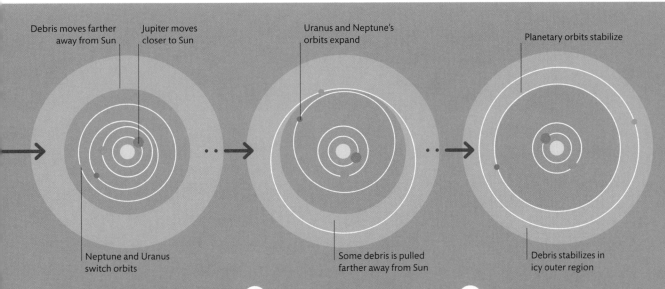

Debris moves farther away from Sun

Jupiter moves closer to Sun

Uranus and Neptune's orbits expand

Planetary orbits stabilize

Neptune and Uranus switch orbits

Some debris is pulled farther away from Sun

Debris stabilizes in icy outer region

2 Planet orbits change
In a theory called the Nice model, Saturn, Uranus, and Neptune are thought to have drifted outward, while Jupiter drifted closer toward the Sun. Uranus and Neptune also switched places with each other.

3 Planets interact with debris
As Uranus and Neptune drifted away from the Sun, they are thought to have carried with them some of their surrounding debris. This brought the debris into the colder outer region of the Solar System.

4 Kuiper Belt stabilizes
Over time, the orbits of the planets and icy objects became stable, creating the Kuiper Belt that exists today. However, some objects are occasionally still disturbed if their orbits bring them too close to Neptune.

Comets

Made up of dust and ice left over from the formation of the planets, comets originate as frozen bodies at the outer edge of the Solar System. In this state, they can be up to tens of miles (km) across. When these objects are knocked out of a regular orbit, they are sent on orbits that bring them close to the Sun. When they approach the Sun, they transform into comets.

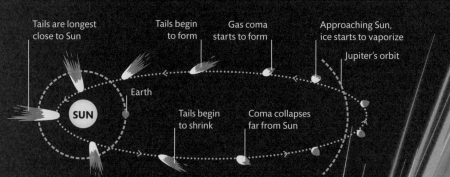

Tails are longest close to Sun

Tails begin to form

Gas coma starts to form

Approaching Sun, ice starts to vaporize

Jupiter's orbit

Earth

SUN

Tails begin to shrink

Coma collapses far from Sun

The life of a comet

When a comet approaches the Sun, ice on its surface vaporizes, creating an atmosphere called a coma and two tails. The coma collapses when the orbit carries the comet far enough away from the Sun and the tails fade.

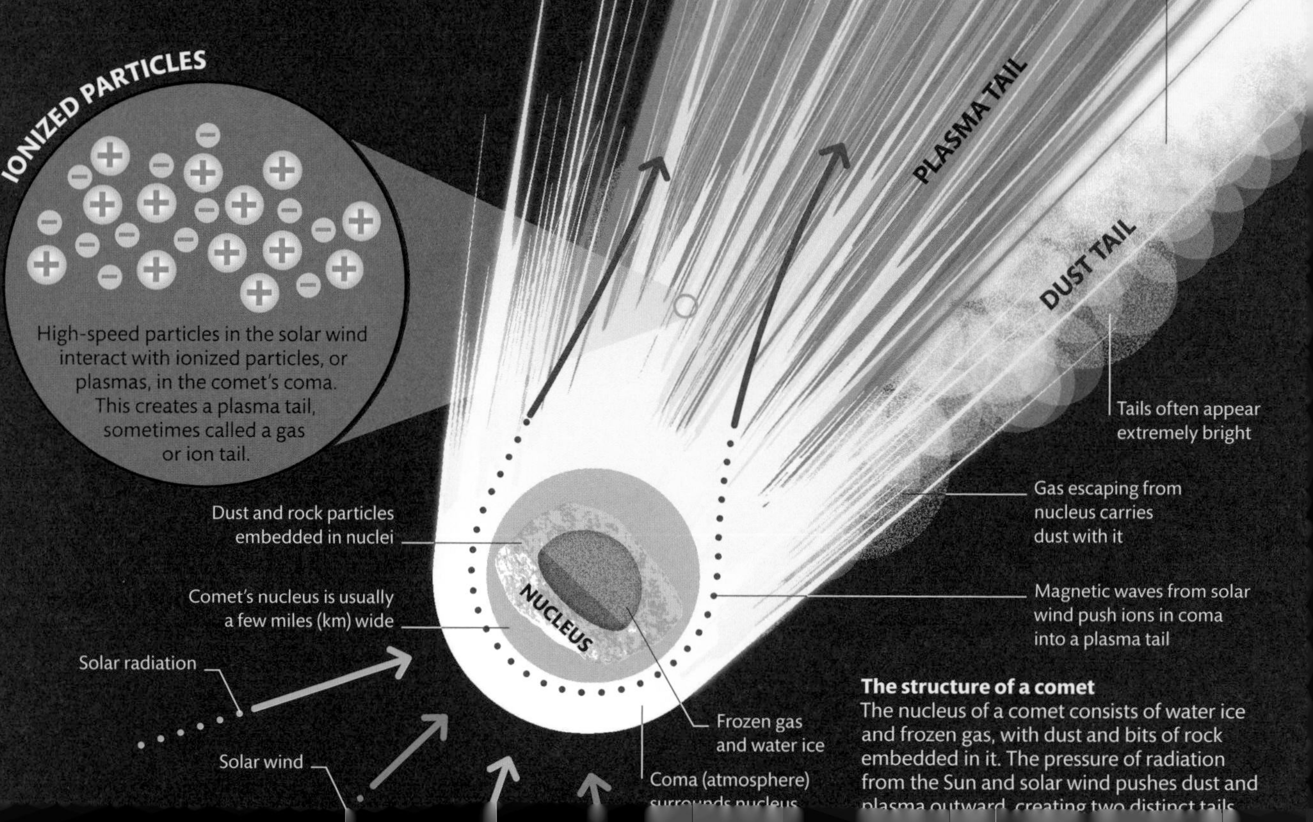

IONIZED PARTICLES

High-speed particles in the solar wind interact with ionized particles, or plasmas, in the comet's coma. This creates a plasma tail, sometimes called a gas or ion tail.

Dust tail becomes curved due to motion of comet along its orbit

PLASMA TAIL

DUST TAIL

Tails often appear extremely bright

Gas escaping from nucleus carries dust with it

Magnetic waves from solar wind push ions in coma into a plasma tail

Dust and rock particles embedded in nuclei

Comet's nucleus is usually a few miles (km) wide

NUCLEUS

Solar radiation

Solar wind

Frozen gas and water ice

Coma (atmosphere) surrounds nucleus

The structure of a comet

The nucleus of a comet consists of water ice and frozen gas, with dust and bits of rock embedded in it. The pressure of radiation from the Sun and solar wind pushes dust and plasma outward, creating two distinct tails.

HOW LARGE IS A COMET'S COMA?

A coma—the atmosphere surrounding the nucleus of a comet—can be thousands of miles (km) across. The comas of some comets are even larger than Earth.

ROGUE PLANETS

Beyond the Oort Cloud, it is possible that there are planet-sized objects, called rogue planets, which do not orbit any star. They might have formed from material that orbited a star and was then ejected, or these rogue planets simply may never have orbited a star.

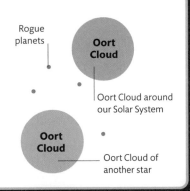

Rogue planets

Oort Cloud

Oort Cloud around our Solar System

Oort Cloud

Oort Cloud of another star

Comet tails can stretch for hundreds of thousands of miles (km)

THE **OORT CLOUD** MIGHT CONTAIN **BILLIONS, OR EVEN TRILLIONS,** OF OBJECTS

Comets and the Oort Cloud

Astronomers think that the Solar System is surrounded by a swarm of icy bodies lying far beyond the Kuiper Belt. Called the Oort Cloud, it is the source of long-period comets that sometimes reach the inner Solar System.

The Oort Cloud

The Oort Cloud is thought to start around 190 billion–470 billion miles (300 billion–750 billion km) from the Sun and end 0.9 trillion–9 trillion miles (1.5 trillion–15 trillion km) from the Sun. This means the outer edge could sit halfway between the Sun and its nearest star. In the Oort Cloud, objects orbit the Sun on paths tilted at all angles, unlike in the Main Belt (see pp.60–61) and the Kuiper Belt (see pp.82–83), where most follow orbits close to the main plane of the Solar System.

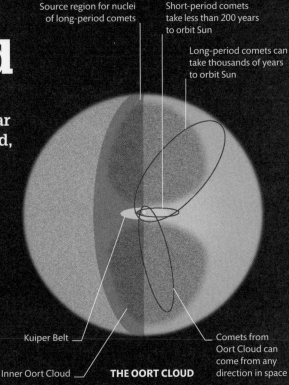

Source region for nuclei of long-period comets

Short-period comets take less than 200 years to orbit Sun

Long-period comets can take thousands of years to orbit Sun

Kuiper Belt

Comets from Oort Cloud can come from any direction in space

Inner Oort Cloud

THE OORT CLOUD

STARS

MAIN-SEQUENCE STAR TYPES

Spectral type	Color	Approximate surface temperature (Kelvin)	Average mass (The Sun = 1)	Average radius (The Sun = 1)	Average luminosity (The Sun = 1)
O	Blue	Over 25,000 K	Over 18	Over 7.4	20,000–1,000,000
B	Blue-white	11,000–25,000 K	3.2–18	2.5–7.4	11,000–20,000
A	White	7,500–11,000 K	1.7–3.2	1.3–2.5	6–80
F	Yellow to white	6,000–7,500 K	1.1–1.7	1.1–1.3	1.3–6
G	Yellow	5,000–6,000 K	0.78–1.10	0.85–1.05	0.40–1.26
K	Orange to red	3,500–5,000 K	0.60–0.78	0.51–0.85	0.07–0.40
M	Red	Under 3,500 K	0.10–0.60	0.13–0.51	0.0008–0.072

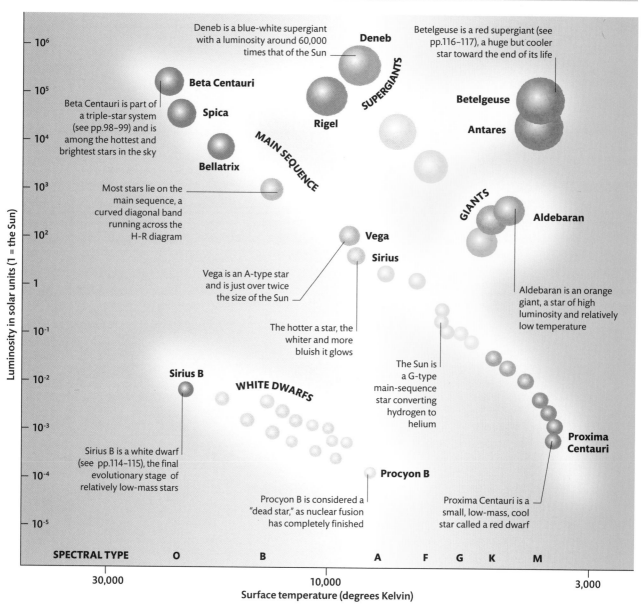

Deneb is a blue-white supergiant with a luminosity around 60,000 times that of the Sun

Betelgeuse is a red supergiant (see pp.116–117), a huge but cooler star toward the end of its life

Beta Centauri is part of a triple-star system (see pp.98–99) and is among the hottest and brightest stars in the sky

Most stars lie on the main sequence, a curved diagonal band running across the H-R diagram

Vega is an A-type star and is just over twice the size of the Sun

The hotter a star, the whiter and more bluish it glows

Aldebaran is an orange giant, a star of high luminosity and relatively low temperature

The Sun is a G-type main-sequence star converting hydrogen to helium

Sirius B is a white dwarf (see pp.114–115), the final evolutionary stage of relatively low-mass stars

Procyon B is considered a "dead star," as nuclear fusion has completely finished

Proxima Centauri is a small, low-mass, cool star called a red dwarf

Luminosity in solar units (1 = the Sun)

10^6 10^5 10^4 10^3 10^2 1 10^{-1} 10^{-2} 10^{-3} 10^{-4} 10^{-5}

SPECTRAL TYPE O B A F G K M

30,000   10,000   3,000

Surface temperature (degrees Kelvin)

Classifying stars

Stars can be classified using the H-R diagram (see left). Those that convert hydrogen into helium through nuclear fusion (see p.90) are known as main-sequence stars. These stars, in the stable middle stages of their lives, are located within a diagonal band in the middle of the H-R diagram. Main-sequence stars are classified into seven groups—O, B, A, F, G, K, and M—according to their spectra, the patterns in the colors of light stars emit caused by the chemical elements they contain. These spectral types run from the hottest O-type down to the coolest M-type stars. Only stars near the ends of their lives, such as white dwarfs and supergiants, fall outside the band. These stars have exhausted their supply of hydrogen and become unstable.

The H-R diagram
This famous chart was named after astronomers Ejnar Hertzsprung and Henry Russell and illustrates the relationship between a star's temperature and luminosity. Stars remain on the curved diagonal main sequence for most of their lives. Low-mass stars are red and at the bottom right. Blue stars at the top left have the highest masses. Giants and supergiants, which have exhausted their hydrogen supply, lie at the top right.

WHAT IS THE BRIGHTEST STAR IN THE NIGHT SKY?

Sirius, also known as the Dog Star, in the constellation Canis Major, is the brightest star, with an apparent magnitude of -1.47.

Types of star

Stars are so far away from us that it is hard to tell how big or even how bright they really are. But astronomers can group them into categories by analyzing their spectra (see pp.26–27), which differ according to a star's size and temperature.

Luminosity and brightness
Luminosity is the energy a star emits each second. The brightness of a star as it appears in our sky is called its apparent magnitude and depends on both the star's luminosity and its distance from Earth. It is measured on a numerical scale in which the brightest stars are given negative or low numbers (the brightest stars have values of around -1) and faint stars are given high numbers. The scale does not work in even-sized steps—a star with a magnitude of 1 is 100 times brighter than a star of magnitude 6.

Luminosity
The size of the white dots represents the true luminosity of stars in the constellation Canis Major. But the stars that radiate the most light may not look like the brightest stars in the night sky to us on Earth if they are far away.

Apparent magnitude
Here, the size shows the apparent brightness of the same Canis Major stars. Notice how Sirius looks much brighter because it is closer, but Aludra, 176,000 times brighter than the Sun, is very dim because it is so distant.

THE **MOST LUMINOUS STARS** EMIT **BILLIONS OF TIMES** MORE LIGHT THAN THE **FAINTEST STARS**

Inside stars

Stars shine because they are heated to enormous temperatures by nuclear reactions. Deep inside, hydrogen nuclei are squeezed together so hard by the gravity of the star that they fuse to form helium nuclei, releasing energy.

A star's energy source

Stars are powered through nuclear fusion, principally through the conversion of hydrogen to helium. We know that this takes place because there is no other way something as massive as a star could generate so much energy over its lifetime. The fusion process in stars releases tiny particles called neutrinos, and on Earth, we can detect neutrinos emanating from the Sun. Studies of vibrations in the Sun also reveal its inner structure, just as earthquakes reveal what is inside the Earth.

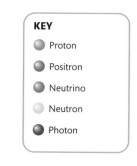

ARE WE MADE OF STARDUST?

Nearly every element in the human body was made in stars over billions of years. The main exceptions are hydrogen and helium, which formed during the Big Bang.

10 BILLION

THE NUMBER OF YEARS IT WILL TAKE THE SUN TO USE UP ALL OF ITS HYDROGEN FUEL

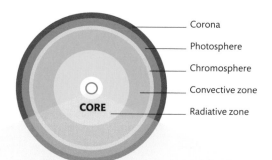

- Corona
- Photosphere
- Chromosphere
- Convective zone
- Radiative zone

CORE

LAYERS OF A STAR SIMILAR TO THE SUN

KEY
- Proton
- Positron
- Neutrino
- Neutron
- Photon

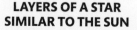

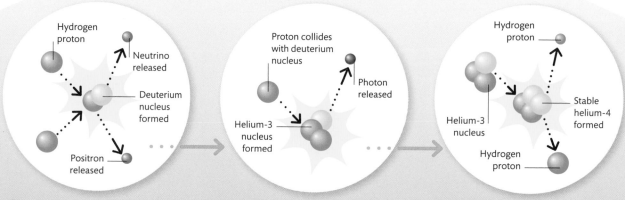

Hydrogen proton
Neutrino released
Deuterium nucleus formed
Positron released

Proton collides with deuterium nucleus
Photon released
Helium-3 nucleus formed

Hydrogen proton
Stable helium-4 formed
Helium-3 nucleus
Hydrogen proton

1 **Protons combine**
Fusion begins when two hydrogen nuclei (protons) join together to form a deuterium nucleus. A positron and a neutrino are released as by-products.

2 **Radiation released**
The deuterium nucleus is hit by another proton, which join to form a helium-3 nucleus. This releases a huge amount of energy in the form of heat and particles called photons.

3 **Helium produced**
The helium-3 nucleus is bombarded by another, creating a helium-4 nucleus. When they join together, they emit two protons, which can start further fusions.

Heat transfer

The layers of stars move heat up and outward mainly by convection and radiation. Convection occurs mostly when radiation is too slow at carrying heat away from the core. In low-mass stars, heat is transferred entirely by convection.

In stars of medium mass, such as the Sun, radiation dominates in the region surrounding the core, but convection takes over in the cooler outer layers, which absorb radiation. In high-mass stars, fusion generates energy so fast that convection dominates around the core.

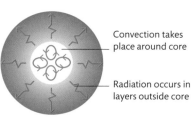

Convection takes place around core

Radiation occurs in layers outside core

OVER 1.5 TIMES THE MASS OF THE SUN

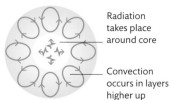

Radiation takes place around core

Convection occurs in layers higher up

0.5–1.5 TIMES THE MASS OF THE SUN

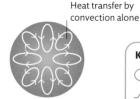

Heat transfer by convection alone

UNDER 0.5 TIMES THE MASS OF THE SUN

KEY

⟳ Convection

∿→ Radiation

Making elements

Most of the lighter natural elements, except hydrogen and helium, were created either by gradual nuclear fusion in stars over their lifetime or when stars suddenly exploded as supernovae. Elements heavier than iron cannot be made in a star's core because iron nuclei cannot be fused. Some of the heavier elements were made in the cores of dying red giants, which do not explode. The rest are believed to come from the violent explosion of two neutron stars merging.

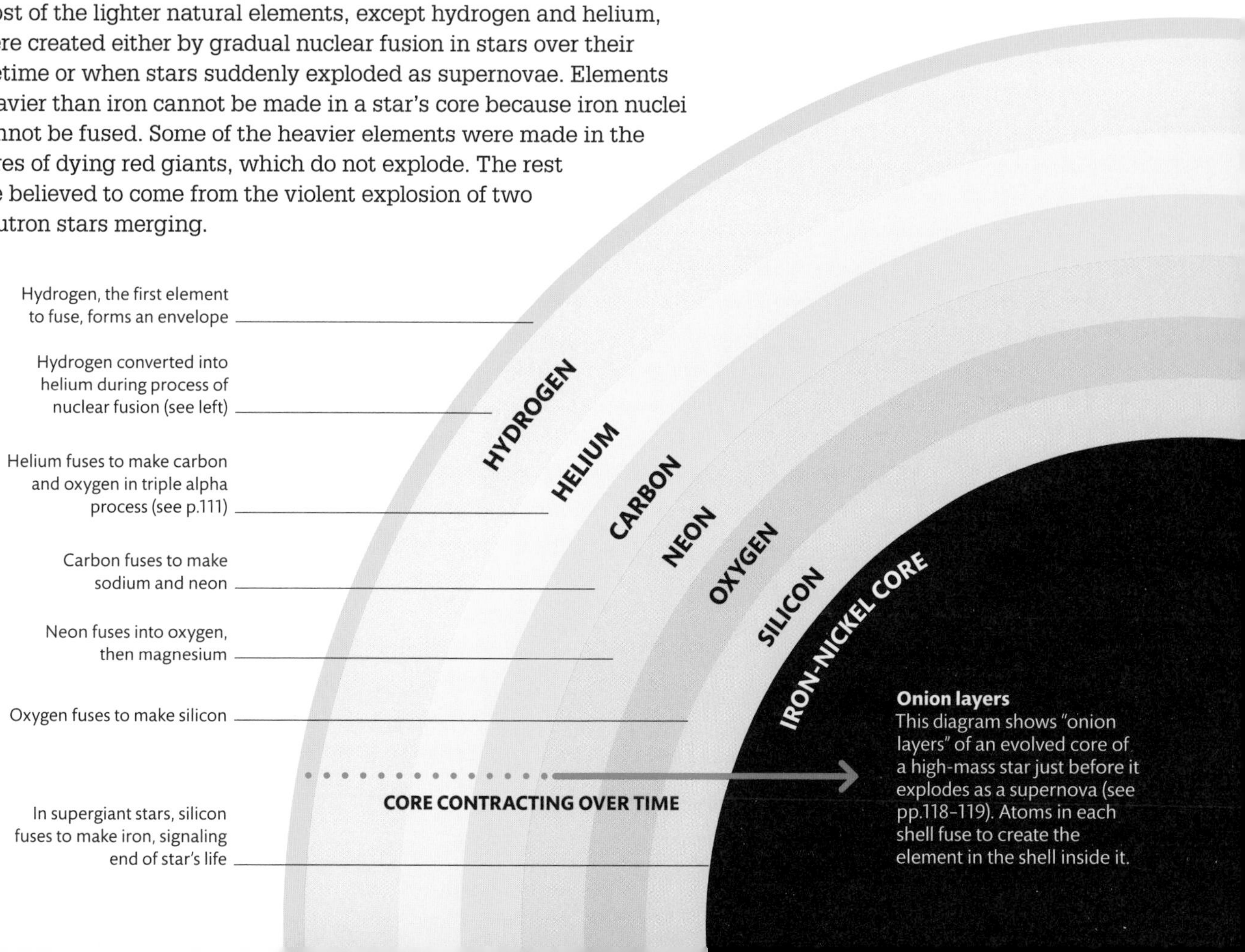

Hydrogen, the first element to fuse, forms an envelope

Hydrogen converted into helium during process of nuclear fusion (see left)

Helium fuses to make carbon and oxygen in triple alpha process (see p.111)

Carbon fuses to make sodium and neon

Neon fuses into oxygen, then magnesium

Oxygen fuses to make silicon

In supergiant stars, silicon fuses to make iron, signaling end of star's life

HYDROGEN

HELIUM

CARBON

NEON

OXYGEN

SILICON

IRON-NICKEL CORE

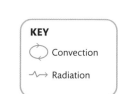

CORE CONTRACTING OVER TIME

Onion layers
This diagram shows "onion layers" of an evolved core of a high-mass star just before it explodes as a supernova (see pp.118–119). Atoms in each shell fuse to create the element in the shell inside it.

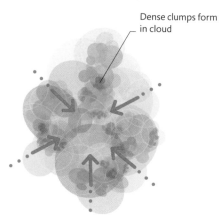

Dense clumps form in cloud

1 **Dense regions form**
The process begins when denser regions form in a space cloud. Molecules in these regions pull in together, creating clumps throughout the cloud. Each one of these clumps may eventually become a star.

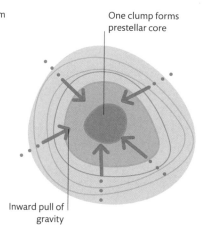

One clump forms prestellar core

Inward pull of gravity

2 **Core collapses**
The core of each clump is denser than the outer parts, so it collapses faster. As a result, it rotates ever faster, conserving angular momentum, like ice skaters pulling their arms in as they spin.

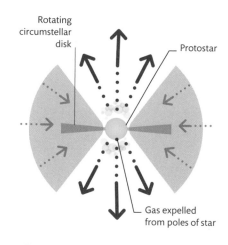

Rotating circumstellar disk

Protostar

Gas expelled from poles of star

3 **Protostar formed**
The prestellar core forms a protostar and is surrounded by a rotating disk of gas and dust. The wider cloud flattens and starts to clear. Some gas is fired out in jets from the poles of the protostar.

Star formation

Stars are continually forming in galaxies all over the Universe. They are born as protostars in vast clouds of gas and dust called giant molecular clouds and go on to evolve into stable main-sequence stars. By studying many stars at different points in their lives, astronomers can determine the stages they undergo.

How a protostar forms

Stars form in dark clouds of gas and dust (see pp.94–95) dense enough to block out light. Starbirth begins when the cloud is disturbed, possibly by shockwaves from a supernova explosion (see pp.118–119), so that clumps of gas and dust begin to pull together under their own gravity. Gravity does the rest.

STAR SIZES AND NUMBERS

There are many more low-mass stars than high-mass stars in the Universe. This is partly because far fewer big stars are born, but also because very big stars have very short lives, so they do not consume fuel and emit light for long. As this graphic shows, for each star of more than 10 solar masses, there are approximately 10 stars of 2–10 solar masses and 50 stars of 0.5–2 solar masses. There are even more red dwarf stars (see pp.88–89)—200 for each star over 10 solar masses.

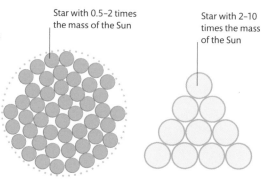

Star with 0.5–2 times the mass of the Sun

Star with 2–10 times the mass of the Sun

Star with over 10 times the mass of the Sun

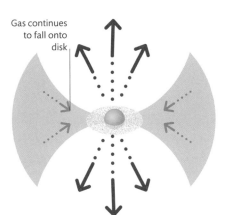

Gas continues to fall onto disk

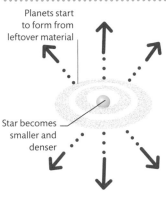

Planets start to form from leftover material

Star becomes smaller and denser

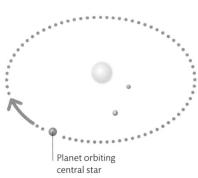

Planet orbiting central star

4 **T Tauri star**
After up to a million years, the central temperature of the protostar reaches 10,800,000°F (6,000,000°C). At this point, hydrogen fusion reactions start and the new star, called a T Tauri star, begins to shine.

5 **Pre-main-sequence star**
After up to 10 million years, the T Tauri star shrinks and grows denser. Material from the disk and the remaining envelope flows into the star or disperses into space. Planets start to form in the disk.

6 **Planetary system created**
The star is now a main-sequence star (see pp.88–89), and planets orbiting the star have fully formed. A planetary system like this typically lives for approximately 10 billion years.

Forces in stars

Once low- and medium-mass stars have begun to fuse hydrogen into helium, they enter the main sequence (see pp.88–89). At this point in the life of a star, the forces inside them—gas pressure emanating from the core and the opposing force of gravity—are balanced. Stars on the main sequence can go on shining steadily for approximately 10 billion years.

Balanced forces
The balance between outward-pushing pressure and inward-pulling gravity in a star is known as hydrostatic equilibrium. It is this balance that keeps a star stable.

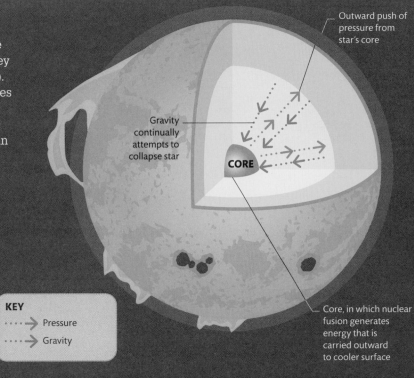

Outward push of pressure from star's core

Gravity continually attempts to collapse star

CORE

Core, in which nuclear fusion generates energy that is carried outward to cooler surface

KEY
····▶ Pressure
····▶ Gravity

WHEN DID THE FIRST STARS APPEAR?

The first stars appeared around 200 million years after the Big Bang. A further billion years passed before galaxies began to proliferate.

STARS ARE THOUGHT **TO FORM** IN THE **UNIVERSE** AT **A RATE** OF ABOUT **150 BILLION A YEAR**

Nebulae

Nebulae are giant clouds in space made of dust and gas. A nebula forms when the sparse material of space clumps together through mutual gravitational attraction. The very densest of nebulae become nurseries for stars.

Diffuse nebulae

Astronomers first noticed nebulae as faint blobs in the night sky back in antiquity, but had no idea what they were. More were spotted after the invention of the telescope, and in 1781 French astronomer Charles Messier included several "diffuse" nebulae in his famous catalogue of astronomical objects. Most nebulae are categorized as "diffuse" because their edges are vague. In turn, diffuse nebulae can be divided into "emission," "reflection," and "dark" nebulae, according to how we see them from Earth. The other types of nebula—planetary nebulae and supernova remnants—are associated with dying and exploding stars.

Types of diffuse nebulae
The key features of the three types of diffuse nebula and the way that they interact with starlight traveling to Earth are shown here.

A NEBULA THE SIZE OF EARTH WOULD HAVE A TOTAL MASS OF ONLY A FEW POUNDS

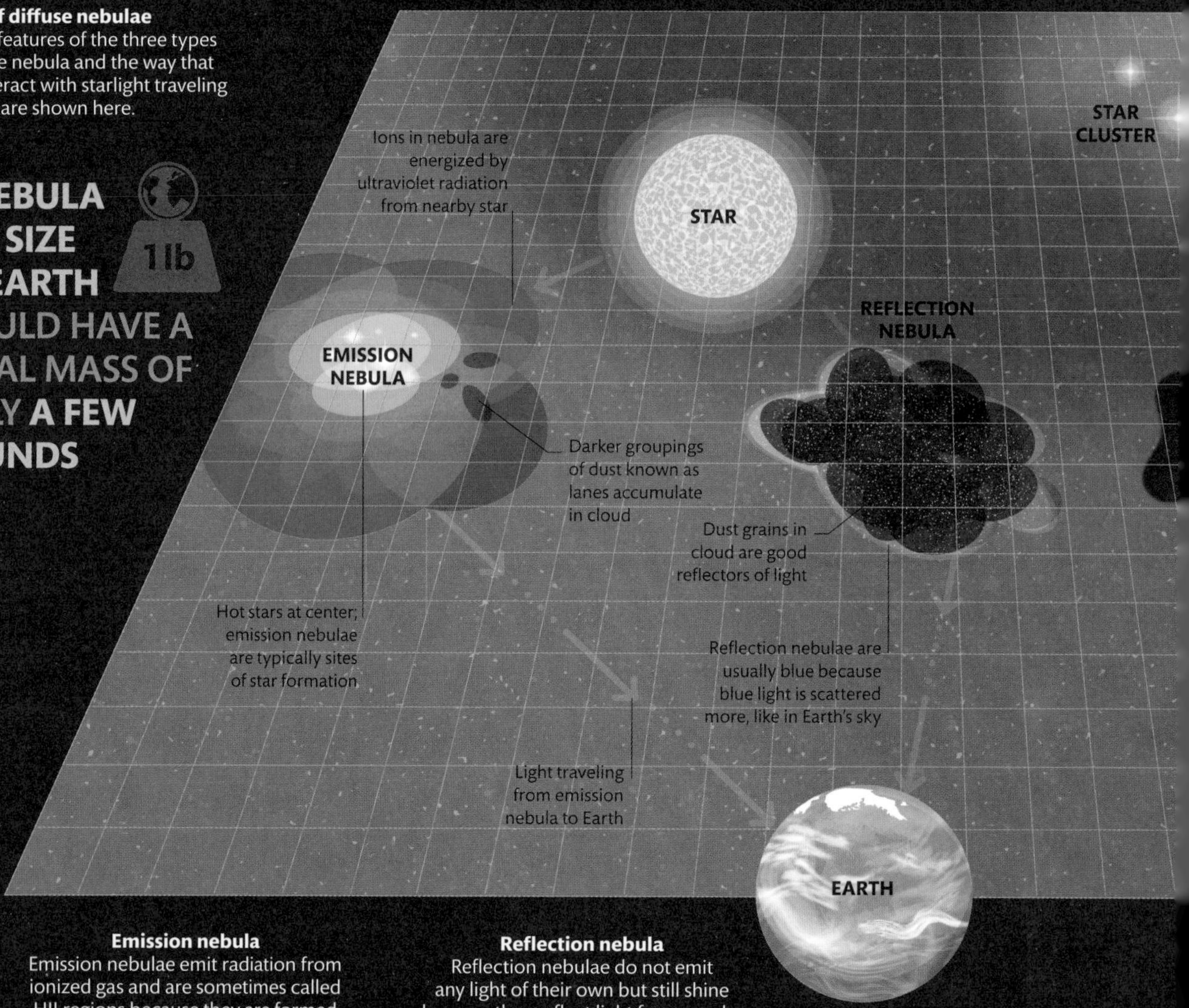

1 lb

Ions in nebula are energized by ultraviolet radiation from nearby star

STAR

STAR CLUSTER

REFLECTION NEBULA

EMISSION NEBULA

Darker groupings of dust known as lanes accumulate in cloud

Dust grains in cloud are good reflectors of light

Hot stars at center; emission nebulae are typically sites of star formation

Reflection nebulae are usually blue because blue light is scattered more, like in Earth's sky

Light traveling from emission nebula to Earth

EARTH

Emission nebula
Emission nebulae emit radiation from ionized gas and are sometimes called HII regions because they are formed mainly of ionized hydrogen.

Reflection nebula
Reflection nebulae do not emit any light of their own but still shine because they reflect light from nearby stars—just like clouds in our own sky.

HOW LARGE CAN A NEBULA GET?

The Tarantula Nebula, located approximately 170,000 light-years from Earth in the Large Magellanic Cloud, stretches for over 1,800 light-years.

Stellar nurseries

Many nebulae are the birthplaces of stars. The most famous is perhaps the Eagle Nebula, where stars are born inside the towering clouds known as the "pillars of creation." These towers, which are each several light-years long, are formed of dense materials that have resisted evaporation by the radiation emitted from nearby young stars.

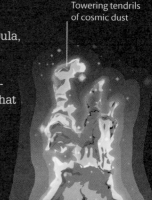

Towering tendrils of cosmic dust

Pillars of creation
This dramatically shaped part of the Eagle Nebula contains hundreds of newly forming stars in its pillars.

Nebulae around dying stars

Planetary nebulae and supernova remnants are also types of nebula and are both created by dying stars. Confusingly, a planetary nebula has nothing to do with planets. It is a shell of gas thrown out by a smaller star as it nears the end of its life. This shell is then ionized by the star's ultraviolet radiation, causing the nebula to glow brightly. A supernova remnant forms when a massive star explodes violently in a supernova, sending a vast cloud of ionized dust and gas out into space.

Blue glow caused by hot helium

Planetary nebula
The Ring Nebula in the constellation Lyra is a remnant of the final stages in the life cycle of a low-mass star.

Pale orange areas show cold dust left from supernova

Supernova remnants
The Crab Nebula in the constellation Taurus is the remnant of a massive star that exploded in 1054 CE.

Light traveling from star cluster to Earth

DARK NEBULA

Dark nebula absorbs light emitted by a star cluster, preventing it from reaching Earth

Dark nebula
Dark, or absorption, nebulae are clouds of dust like reflection nebulae; they only look different because they block out the light from behind.

FALSE-COLOR IMAGING

Objects in space, including nebulae and galaxies, often emit radiation that our eyes cannot detect because it lies outside the visible spectrum. To make pictures of these objects, astronomers use software to assign colors that we can see to the various intensities of radiation they have measured. These pictures are called false-color images.

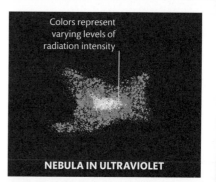

Colors represent varying levels of radiation intensity

NEBULA IN ULTRAVIOLET

Star clusters

Some stars belong to groups called clusters. Open clusters are loose groups of young stars formed from the same cloud of gases and dust. Globular clusters are giant balls of ancient stars.

Types of cluster

Open clusters are mostly just a few tens of millions of years old. The stars are often slightly bluish because they contain remnants of the original cloud. Globular clusters are almost as old as the universe, and the gas and giant stars are long since gone. They can include groups of thousands or millions of stars, bound together by gravity.

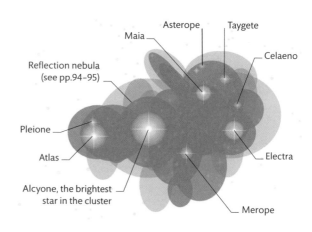

Asterope Taygete

Maia

Celaeno

Reflection nebula (see pp.94–95)

Pleione

Atlas

Electra

Alcyone, the brightest star in the cluster

Merope

Open cluster

The Pleiades is an open cluster of around 3,000 stars that is visible to the naked eye. It is less than 100 million years old and dominated by nine young, bright blue giant stars. The brightest stars of the Pleiades are named after the Seven Sisters of Greek mythology, along with their parents Atlas and Pleione.

THE PLEIADES STAR CLUSTER APPEARS ON THE NEBRA SKY DISK, DATING FROM 1600 BCE

HOW DO WE WORK OUT THE AGE OF STAR CLUSTERS?

Astronomers can tell the age of a star cluster from its mix of stars of different kinds. The older a cluster, the greater the number of stars that have evolved into giants.

How an open cluster develops

Stars are born in large clouds of molecular gas, so they inevitably form in clusters, as these clouds contain the matter needed to create thousands of stars. Clusters contain stars of all types, from relatively cool red dwarfs to massive blue giants. Most clusters last only a few hundred million years, as the biggest stars die out and many loosely bound small stars are pulled away by

Large molecular cloud formed from particles of interstellar gas and dust

Denser parts of the cloud start to collapse inward, pulled by their own gravity

1 Stars are born
Very young stars, which are known as protostars, form where dense concentrations of gas collapse under gravity in a molecular cloud. This can be triggered by the shockwave

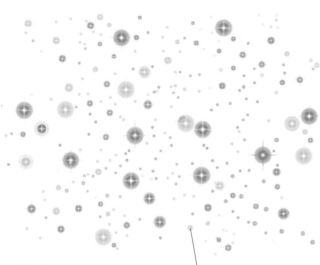

Globular cluster
The stars in the vast Omega Centauri cluster are over 10 billion years old. It is over 16,000 light-years away, yet its 10 million stars shine so brightly together that the cluster is visible to the naked eye, looking like a single star.

Unusually for globular clusters, Omega Centauri includes stars of various ages, most of which are small yellow and white stars

BLUE STRAGGLERS

Globular clusters are mostly so ancient that they should not contain young blue stars—yet some of them do. "Blue stragglers" are thought to form because stars are so closely packed near the center of the cluster that old red stars can collide. When they do, the collision forms a new, high-mass blue star and pumps hydrogen into its core.

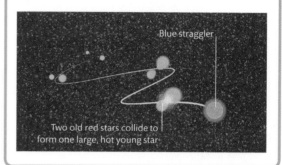

Blue straggler

Two old red stars collide to form one large, hot young star

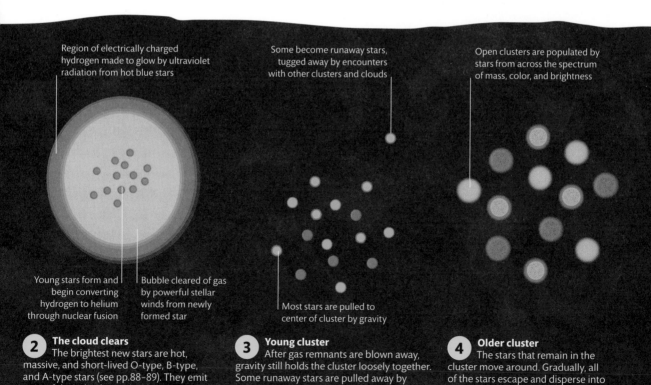

Region of electrically charged hydrogen made to glow by ultraviolet radiation from hot blue stars

Some become runaway stars, tugged away by encounters with other clusters and clouds

Open clusters are populated by stars from across the spectrum of mass, color, and brightness

Young stars form and begin converting hydrogen to helium through nuclear fusion

Bubble cleared of gas by powerful stellar winds from newly formed star

Most stars are pulled to center of cluster by gravity

2 **The cloud clears**
The brightest new stars are hot, massive, and short-lived O-type, B-type, and A-type stars (see pp.88–89). They emit powerful winds of particles that clear away

3 **Young cluster**
After gas remnants are blown away, gravity still holds the cluster loosely together. Some runaway stars are pulled away by gravitational interactions with other clusters

4 **Older cluster**
The stars that remain in the cluster move around. Gradually, all of the stars escape and disperse into space over a period of a few

1 How a Cepheid pulsates

Some stars pulsate because radiated energy is continually trapped, then released by the helium in a particular layer of the star. This occurs because the helium atoms change between two different electrically charged states.

2 Helium becomes transparent

As the helium atoms heat up, they lose one of their two electrons. This makes the gas more transparent to radiation, allowing energy to escape.

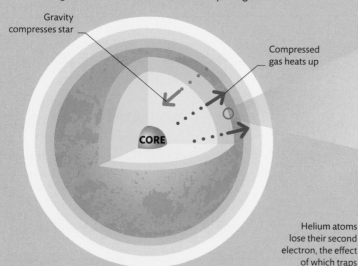

Gravity compresses star

Compressed gas heats up

CORE

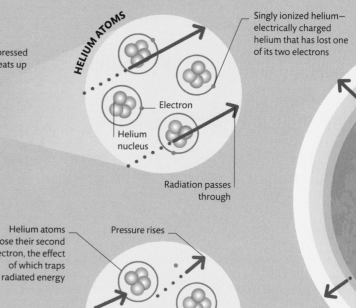

HELIUM ATOMS

Singly ionized helium—electrically charged helium that has lost one of its two electrons

Electron

Helium nucleus

Radiation passes through

Helium atoms lose their second electron, the effect of which traps radiated energy

Pressure rises

Electron moves freely

Variable stars

A variable star is a star that changes in brightness on a timescale ranging from fractions of a second to years. With extrinsic variable stars, the variation is an illusion caused by the star's rotation or another star or planet moving in front of it. With intrinsic variable stars, such as Cepheids (shown below), the change is due to physical changes in the star itself.

KEY

······→ Pressure ······→ Gravity ······→ Radiation

3 Helium becomes opaque

The helium atoms lose their remaining electron, which makes the gas more opaque. This means that the energy traveling from the star's core is trapped, so pressure inside the star rises and the star swells.

Cepheid variables

Cepheids are a type of variable star that exhibit a relationship between their period (the time it takes to brighten, dim, and brighten again) and their luminosity. The brighter a Cepheid is, the longer its period, so timing its period shows how bright it is. Comparing the period to the star's apparent brightness means that it is possible to work out exactly how far it is away from Earth.

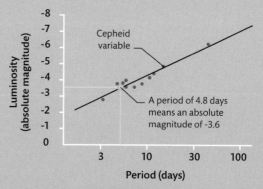

Luminosity (absolute magnitude)

Cepheid variable

A period of 4.8 days means an absolute magnitude of -3.6

Period (days)

Period-luminosity relationship

When you know the period of a Cepheid, you can use a period-luminosity chart to work out its absolute magnitude. An equation is then used to calculate its distance from Earth.

UP TO **85 PERCENT** OF ALL STARS ARE PART OF **MULTIPLE-STAR SYSTEMS**

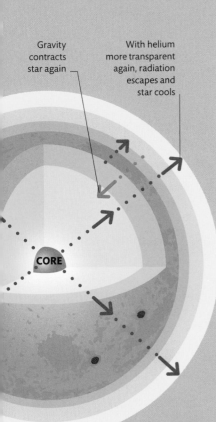

Gravity contracts star again

With helium more transparent again, radiation escapes and star cools

CORE

4 Radiation released
As the star expands, the helium cools. Helium atoms revert back to their singly ionized state, which allows radiation to escape. Pressure inside the star drops and gravity pulls the star in again, recompressing the gas.

HOW MANY STARS CAN EXIST IN A SYSTEM?

The star systems AR Cassiopeiae and Nu Scorpii are the only known examples of septenary star systems (systems of seven stars). There are several sextenary systems.

Multiple and variable stars

It may look as if all the points of light in the sky are lone stars like our Sun. In fact, over half are in pairs called binaries and two-thirds of the rest are in even bigger groups. More than 150,000 stars are variable stars that fluctuate in brightness.

Binary stars

Binaries are two stars orbiting a common center of mass, known as a barycenter. The brighter star of the two is called the primary. Multiple groups include three or more stars circling around each other in complex patterns. Some binary stars are too far apart to have much of a gravitational effect on each other. Others are so close that one star can draw mass from the other—sometimes so much that it becomes a black hole (see pp.122–123).

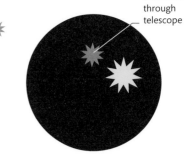

Optical doubles
Two stars that are not together, like true binaries, but just in the same line of sight from Earth are called optical doubles. They may not look like it, but the two stars are often at vast distances from each other.

STAR B
STAR A
EARTH

Stars seen through telescope

FROM SPACE

FROM EARTH

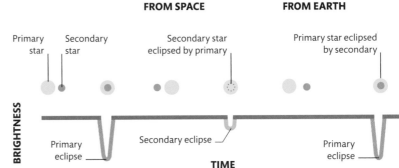

Primary star   Secondary star

Secondary star eclipsed by primary

Primary star eclipsed by secondary

BRIGHTNESS

Primary eclipse

Secondary eclipse

Primary eclipse

TIME

Eclipsing binary stars
These are two stars whose orbits are in line as seen from Earth, so that one regularly passes in front of the other, causing the combined brightness to dip. This repeated eclipse can give the illusion that the star is flashing on and off.

Between the stars

The space between the stars, known as the interstellar medium (ISM), contains gas and dust that plays an important role in the evolution of stars. Within the ISM are distinct regions characterized by differences in temperature, density, and electric charge.

In the densest diffuse clouds, known as HI regions, the constituent hydrogen atoms are entirely neutral; temperatures in these regions range from -280° F (-170° C) to 1,340° F (730° C)

Interstellar gas

Around 99 percent of the ISM is gas—mostly hydrogen. On average, each cubic centimeter of the ISM is occupied by only one atom (compared to 30 million trillion molecules per cubic cm in the air we breathe). But over the vastness of space, that is enough to form visible clouds. These are either cold clouds of neutral hydrogen or hot clouds of charged hydrogen near young stars. Helium is the second most common element, but many others are also found in very small quantities as individual atoms or in molecules.

1 Clouds form
Interstellar clouds form from the gas and dust particles expelled by dying red giants (see pp.110–111). Diffuse clouds are the least dense of these clouds, dominated by neutral or charged (ionized) hydrogen.

2 Dense regions form
Dust grains and gas particles in diffuse clouds may gather together because of their mutual gravitational attraction.

HI REGION

DIFFUSE CLOUD

COLD INTERSTELLAR MEDIUM

In coolest parts of cold interstellar medium, temperatures are as low as -440°F (-260°C)

Some regions of the ISM are heated to temperatures up to about 18,000°F (10,000°C)

CORONAL INTERSTELLAR GAS

Many galaxies are surrounded by a vast, tenuous halo or corona of hot ionized gas

6 Red giant
An aging medium-mass star uses up its fuel and collapses, scattering dust and gas to form new clouds. On average, a third of the matter drawn into stars goes back into interstellar space.

RED GIANT

WARM INTERSTELLAR MEDIUM

6 Supernova
An aging high-mass star becomes a supergiant, which eventually goes supernova (see pp.118–119). The debris from the explosion adds new material to the ISM.

AROUND **15 PERCENT** OF ALL THE VISIBLE MATTER IN THE **MILKY WAY** IS **INTERSTELLAR GAS AND DUST**

PLANET ORBITING STAR

5 Protoplanetary system
When a new star forms, dust collects in a disk rotating around the star, then clumps together to form planets.

Proton and electron spin in same direction

PROTON

ELECTRON

Electron spontaneously spins in other direction

21-cm-long waves of radiation emitted when electrons reverse their direction of spin; these waves can be detected by radio telescopes

Detecting cold clouds
Neutral hydrogen atoms (protons) in HI regions can be detected when their electrons spontaneously reverse their direction of spin.

Interstellar dust
Interstellar dust is mostly atomic soot belched out by stars. It is composed of tiny grains containing silicates (compounds of oxygen and silicon), carbon, ice, and iron compounds. These irregularly shaped microscopic grains are 0.01–0.1 micrometers (millionths of a meter) in diameter and are warmer than the surrounding gas. Interstellar dust accounts for around 1 percent of the total mass of the ISM.

MOLECULAR CLOUD

PRESTELLAR CORE

3 Forming clumps
Molecular clouds are much smaller and denser than diffuse clouds. Within them, hydrogen forms molecules and dust and gas combine to produce clumps that form prestellar cores.

Star emitting blue and red wavelengths of light

Red light is not scattered as much by dust, so more of it reaches observer

STAR

INTERSTELLAR CLOUD

OBSERVER

Reddening effect
Blue light gets scattered much more than red light by interstellar dust, so stars often appear reddish.

Dust particles roughly the same size as wavelength of blue light absorb and scatter blue light more than red light

4 Star formation
In places, clumps gather together enough material and grow big enough to create the interior pressure needed to form stars.

In a cloud where stars form, called an HII region, heat from stars ionizes much of cloud's hydrogen; electrons emit light as they shift energy levels, making cloud glow

NOBLE COMPOUNDS

It was thought that some gases, called noble gases, could not combine with other elements. But in the extreme conditions of the ISM, the impossible can happen. Helium has been detected joining with hydrogen, and argon can combine with hydrogen to form the compound argonium.

HII REGION

NEW STAR

IS INTERSTELLAR SPACE A VACUUM?

In parts, the ISM is the closest thing to a vacuum. Densities in coronal interstellar gas are far lower than laboratory vacuums on Earth, but nowhere in space is totally empty.

Argonium, made from an argon atom and a proton, can form in ISM

ARGON NUCLEUS

Hydrogen nucleus, or a proton

The ISM cycle
Stars are formed out of the ISM. Then, when they die, much of their matter, including new elements created inside stars and in stellar explosions, is expelled back into the ISM to start the cycle again.

Exoplanets

Our Sun is not the only star to be orbited by other planets. Since 1995, when the first exoplanets were discovered, over 4,000 more have been found. With ongoing missions dedicated to the search, the total is increasing all the time.

How planets form

There are two theories about how planets form: one to do with building from the top down, the other from the bottom up. In the bottom-up theory, core accretion, planets slowly form through collisions between increasingly large pieces of debris in the disk of gas and dust surrounding a young star. In the top-down theory, disk instability, giant planets can result when large clumps of gas form in the disk of material surrounding a young star.

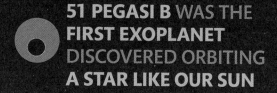

51 PEGASI B WAS THE **FIRST EXOPLANET** DISCOVERED ORBITING **A STAR LIKE OUR SUN**

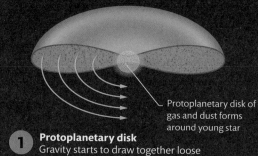

CORE ACCRETION THEORY

Protoplanetary disk of gas and dust

Central star, typically a few million years old

1 Dust collision
Swirling dust grains in a protoplanetary disk collide, forming bigger and bigger clumps. This process creates mini planets called planetesimals.

DISK INSTABILITY THEORY

Protoplanetary disk of gas and dust forms around young star

1 Protoplanetary disk
Gravity starts to draw together loose clumps of gas in the cooler, outer parts of the protoplanetary disk.

Types of exoplanet

As astronomers learn more about exoplanets, they group them into loose categories, comparing them to the planets in the Solar System and to Earth in particular. Some categories depend on a planet's mass, such as super-Earths and mega-Earths. Some of the smaller exoplanets may be covered in oceans and are known as water worlds. Other categories depend on how closely the planet orbits the star. Hot Jupiters and hot Neptunes are gas giants in tight, fast orbits around their stars. Exo-Earths, such as TOI 700d, discovered in 2020, are possibly the most interesting due to their potential habitability.

Hot Jupiter
These gas giants have a similar mass to Jupiter but are much closer to their host stars and are therefore much hotter.

Chthonian planet
This is the solid remnant core of a gas giant. The atmosphere has been stripped away due to its close proximity to its star.

Mega-Earth
First used for Kepler-10c in 2014, "mega-Earth" refers to a rocky planet with at least 10 times Earth's mass.

Super-Earth
These can be up to 10 times the size of Earth. The first super-Earth with water in its skies was found in 2019.

Water world
A terrestrial planet with surface water or a below-surface ocean; the first, GJ 1214B, was discovered in 2012.

Exo-Earth
This is a planet with a size and mass similar to Earth and located within the habitable zone of its star.

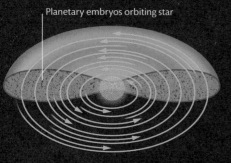

Planetary embryos orbiting star

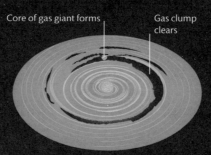

Rocky planets begin to form

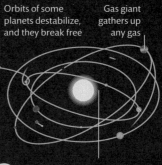

Orbits of some planets destabilize, and they break free

Gas giant gathers up any gas

2 **Planetary embryos form**
The planetesimals grow to form the embryos of planets and begin to move in orbits around the central star.

3 **Rocky planets form**
Close to the star, heavier metallic elements condense and violent collisions can lead to the creation of rocky planets.

4 **Gas giants created**
Farther out, cooler temperatures allow hydrogen and helium to condense to form gas giants.

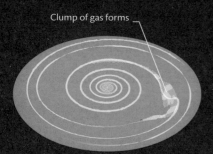

Clump of gas forms

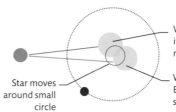

Core of gas giant forms

Gas clump clears

Planet sweeps a wide gap

2 **Separating out**
A clump containing enough gas to form a giant planet cools rapidly. It shrinks and becomes denser.

3 **Core forming**
Dust grains are drawn in by the gravity of the massive gas clump. They fall to the center to form the core of a giant planet.

4 **Planetary sweeper**
The new planet sweeps through the disk, clearing a wide path and growing as it gathers in gas and dust on its way.

Detecting exoplanets

Exoplanets are tiny compared to their parent star and are often hidden by the star's glare, because they emit no light of their own. Only a few giant exoplanets have been photographed directly, a technique called direct imaging. Most are detected indirectly using methods called transit photometry and radial velocity. Just under 100 exoplanets have been discovered by a process called gravitational microlensing, which involves a chance alignment of a nearby star with planets and a distant star. The exoplanets reveal themselves as they bend the distant star's light a bit like a lens.

Radial velocity method
When a large planet orbits a star, its gravitational pull causes the star to move around a small circle so that the light waves it emits change color.

When star moves toward Earth, its light waves are shortened, making them bluer

Star moves around small circle

When star moves away from Earth, its light waves are stretched, making them redder

RADIAL VELOCITY METHOD

Transit photometry method
When a planet passes in front of the star it orbits, we cannot see the planet directly, but the star dims slightly, which can be measured.

Light output

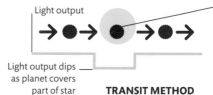

Intensity of star's light drops measurably as planet passes in front of it, in a kind of eclipse

Light output dips as planet covers part of star

TRANSIT METHOD

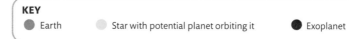

KEY
● Earth   ● Star with potential planet orbiting it   ● Exoplanet

Finding other Earths

Ever since astronomers discovered the first exoplanet in 1995, they have been hunting for planets similar to Earth. The search centers on areas around stars known as habitable zones, where conditions may be right for life. So far, more than 50 planets have been discovered in habitable zones.

The Goldilocks zone

Water is essential for life—so the habitable zone around every star is the zone where the temperature is right to maintain liquid surface water. This zone is sometimes called the Goldilocks zone, because it is neither too hot nor too cold, like the porridge that Goldilocks favored in the fairy tale. If the planet is too hot, water will boil away; if it is too cold, then water will freeze. In a system containing a big, hot star, the habitable zone is much farther out than it is in a system with a small, cooler star.

CAN EXOPLANETS ORBIT MORE THAN ONE STAR?

Astronomers have found over 200 double stars with planets. Kepler-64 was the first quadruple-star system found with a planet orbiting two of the stars.

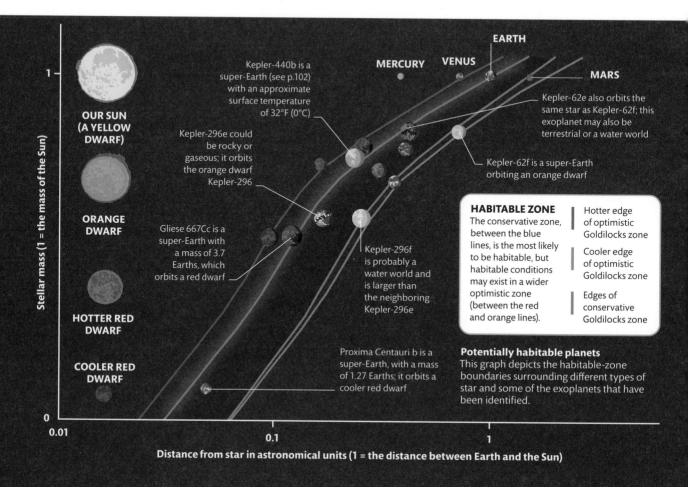

OUR SUN (A YELLOW DWARF)

ORANGE DWARF

HOTTER RED DWARF

COOLER RED DWARF

Stellar mass (1 = the mass of the Sun)

Kepler-440b is a super-Earth (see p.102) with an approximate surface temperature of 32°F (0°C)

Kepler-296e could be rocky or gaseous; it orbits the orange dwarf Kepler-296

Gliese 667Cc is a super-Earth with a mass of 3.7 Earths, which orbits a red dwarf

Kepler-296f is probably a water world and is larger than the neighboring Kepler-296e

Proxima Centauri b is a super-Earth, with a mass of 1.27 Earths; it orbits a cooler red dwarf

MERCURY   VENUS   EARTH   MARS

Kepler-62e also orbits the same star as Kepler-62f; this exoplanet may also be terrestrial or a water world

Kepler-62f is a super-Earth orbiting an orange dwarf

HABITABLE ZONE
The conservative zone, between the blue lines, is the most likely to be habitable, but habitable conditions may exist in a wider optimistic zone (between the red and orange lines).

- Hotter edge of optimistic Goldilocks zone
- Cooler edge of optimistic Goldilocks zone
- Edges of conservative Goldilocks zone

Potentially habitable planets
This graph depicts the habitable-zone boundaries surrounding different types of star and some of the exoplanets that have been identified.

0.01   0.1   1

Distance from star in astronomical units (1 = the distance between Earth and the Sun)

What makes a planet habitable?

When searching for potentially habitable planets, astronomers look mostly for rocky planets, like Earth. Once a likely exoplanet has been identified, research efforts focus on establishing other factors that might make it a prime candidate for life, such as a moderate surface temperature and liquid surface water. NASA's Transiting Exoplanet Survey Satellite (TESS), which launched in 2018, scans the sky for planets in the habitable zone. It is the successor to the Kepler Space Telescope (see pp.186–187), which detected over 2,600 exoplanets.

Temperature
This must be moderate to keep water liquid. If it is too cold, chemical reactions may be too slow to sustain life.

Surface water
Liquid surface water would make life very likely, but it is possible that water underground may support life.

Stable sun
The nearest star must remain stable and shine steadily in order for life to evolve on a rocky planet.

Elements
The building blocks of life, including carbon, oxygen, and nitrogen, need to be present.

Spin and tilt
A tilted spin axis stops extremes of temperature. Planets that do not spin can be very hot on the side facing the star.

Atmosphere
An atmosphere traps warmth, shields the surface from harmful radiation, and stops gases from escaping.

Molten core
A molten core can create a magnetic field that shields life from some of the radiation coming from outer space.

Sufficient mass
Without sufficient mass, a planet will not have enough gravity to hold on to its water or its atmosphere.

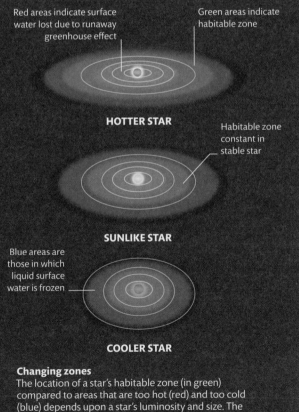

Red areas indicate surface water lost due to runaway greenhouse effect

Green areas indicate habitable zone

HOTTER STAR

Habitable zone constant in stable star

SUNLIKE STAR

Blue areas are those in which liquid surface water is frozen

COOLER STAR

Changing zones
The location of a star's habitable zone (in green) compared to areas that are too hot (red) and too cold (blue) depends upon a star's luminosity and size. The edges of habitable zones change as stars age, especially as they reach the ends of their lives.

THE **KEPLER-90** SYSTEM CONTAINS **8 EXOPLANETS,** THE SAME NUMBER AS IN OUR **SOLAR SYSTEM**

THE MOST EARTHLIKE PLANET

The exoplanet Kepler-1649c is 300 light-years from Earth. NASA described it as the "most similar to Earth in size and estimated temperature" out of the thousands of exoplanets discovered by the Kepler space telescope. It was discovered on April 15, 2020.

Earth

Kepler-1649c

The four ingredients
There are thought to be four ingredients that make life possible: water, energy, organic chemicals, and time. Without these, life is unlikely to be supported in other worlds.

Chemical reactions
Almost all the processes that make up life on Earth involve chemical reactions—and most of those reactions require a liquid to break down substances so they can move and interact freely. The best and most abundant liquid for this purpose is water.

Energy input
No life form can survive without energy. On Earth, sunlight is the key energy input. But in the early days of Earth, it may have been lightning triggered by volcanic eruptions that provided the vital spark.

Complex molecules condense on side of flask

SODIUM CHLORIDE

Lattice structure of salt, comprising positively charged sodium and negatively charged chloride ions

Boiling water, methane, ammonia, and hydrogen

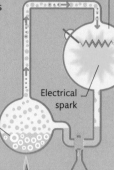

Electrical spark

1　Dissolving salt
When dissolving sodium chloride (salt), water molecules pull the sodium and chloride ions apart, breaking their bonds.

Molecules collected

Chloride ion

Water molecule, made of two hydrogen atoms bonded to oxygen atom

Sodium ion

The Miller-Urey test
In 1952, an experiment simulated lightning to prove that, given enough energy, complex organic molecules can form from simple inorganic materials.

TIME

Sufficient time
The journey from single-celled organisms to complex life requires a time period of billions of years.

2　Solution formed
After the bonds are pulled apart, the sodium and chloride ions are surrounded by water molecules to form a solution.

Amino acid glycine, as found on a comet by the Rosetta probe in 2016 (see pp.194–195)

HYDROGEN

OXYGEN

NITROGEN　CARBON

GLYCINE

Carbon-based chemicals
Organic chemicals are the basis of life on Earth. Yet these molecules, including complex amino acids, are abundant elsewhere in the Universe, too, detectable in huge quantities in nebulae and identified in meteorites that hit Earth.

1　Inorganic ingredients
As on Earth, complex mixtures of gases in a planet's atmosphere could provide the sources of life's principal elements: carbon, hydrogen, oxygen, and nitrogen.

2　Simple organic molecules
Charged with enough energy, atoms of carbon, hydrogen, and other elements can combine to form the organic molecules (some carbon compounds) needed for life, such as amino acids.

LIFE ON EARTH MAY STRETCH BACK 4.3 BILLION YEARS

ORGANIC MOLECULES

Is there life in the Universe?

Life on Earth may be unique, but most scientists think this is unlikely. The Universe is so vast that it is possible that the conditions that created life on Earth could also exist elsewhere.

Ingredients for life

Scientists who search for life in space, known as astrobiologists, believe that there are four key ingredients for life to begin: water, organic molecules, energy, and time. Water is essential for life, because it dissolves chemical nutrients for organisms to eat, transports vital chemicals inside cells, and enables cells to remove waste. The right chemical ingredients are also needed to make life possible. Carbon is top of the list, because its unique ability to form bonds with itself and other elements enables it to form the complex molecules crucial to life, such as proteins and carbohydrates.

ENCELADUS

Since discovering extremophiles, astrobiologists have renewed their search for signs of life in more extreme places in the Solar System, including Saturn's moon Enceladus. In 2011, plumes of water vapor containing salts, methane, and complex organic molecules were found erupting through its icy surface from oceans beneath.

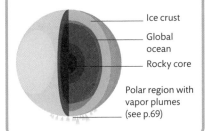

- Ice crust
- Global ocean
- Rocky core
- Polar region with vapor plumes (see p.69)

Active state
In its active state, a tardigrade can eat, grow, move, fight, and reproduce.

Anoxybiosis
If the water in its environment loses oxygen, a tardigrade puffs up and becomes turgid.

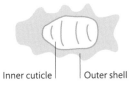

Inner cuticle | Outer shell

Encystment
To adapt to harsh environments, it makes itself a tough outer shell and retracts within a cuticle.

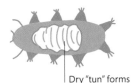

Dry "tun" forms

Anhydrobiosis
In very dry conditions, it shrivels into a dry ball (tun) and survives by consuming special proteins.

Extremophiles

On Earth, microbes have been discovered in hostile places such as in boiling water around vents in the ocean floor. These extremophiles, life forms that thrive in extreme conditions, suggest that life can develop in a huge range of environments. The tardigrade, an aquatic micro-animal, can enter various states to adapt to its surroundings (see left). In one of these states, anhydrobiosis, a tardigrade stops its metabolism and shrivels up. In this state, a tardigrade can even survive in the harsh conditions of outer space.

How stars age

Most stars seem unchanging, but over billions of years, they are born, age, and finally die. We can see examples of nearly all the different stages of stellar evolution by studying the stars in our galaxy and beyond.

A star's life story

After a new star enters the main sequence (see pp.88–89), it steadily converts hydrogen to helium through nuclear fusion in its core. This can take place for billions of years, with outward pressure from fusion balancing the inward force of gravity. When a star has used up all the hydrogen in its core, it enters the last stages of its life. What happens then depends on a star's mass. Low-mass stars shrink and are thought to fade into black dwarfs; medium-mass stars expand into red giants, then collapse as white dwarfs; and high-mass stars become supergiants, then explode in supernovae.

11,000,000°F
(6,000,000°C) IS THE APPROXIMATE TEMPERATURE AT WHICH NUCLEAR FUSION BEGINS IN A STAR'S CORE

HOW LONG DOES A STAR SPEND ON THE MAIN SEQUENCE?

Stars spend 90 percent of their lives on the main sequence, converting hydrogen to helium. The final stages of their lives occur relatively quickly.

Red dwarfs are very low-mass stars and the smallest, coolest stars on the main sequence

1 Low-mass star
The lower the mass of a star, the longer it stays on the main sequence before entering its final stages.

Star on the main sequence

Medium-mass star, which has almost exhausted the hydrogen in its core

Main sequence
A star enters the main sequence once the hydrogen fusion that makes it shine begins. Its life after that can go one of three ways, depending on its initial mass.

1 Medium-mass star
Stars such as the Sun burn slowly over around 10 billion years before they use up the hydrogen in their cores.

OLDER THAN THE UNIVERSE?

HD 140283, described as the "Methuselah" star, is one of the Universe's oldest known stars. In 2000, scientists calculated its age as 16 billion years, but that was impossible because the Universe itself is only 13.8 billion years old. In 2019, the star's age was recalculated as about 14.5 billion years, but with a margin of error of 800 million years. Whatever its precise age, HD 140283 is very old indeed.

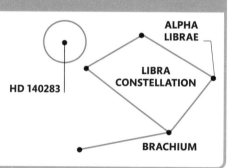

HD 140283

ALPHA LIBRAE

LIBRA CONSTELLATION

BRACHIUM

1 High-mass star
The most massive stars burn bright and fast, some for as little as 20 million years.

Star starts to decrease in size
as inward-pulling gravity
is now stronger than
outward-pushing pressure

Small, dim star
gradually fades

A low-mass star might last
80 billion years before
collapsing to form
hypothetical black dwarf

2 **Fusion ceases**
All the hydrogen in
the star's core has been
used up, so it converts
the hydrogen in its
atmosphere to helium
and starts to collapse.

3 **Shrinking down**
The star cannot
generate enough heat in
its core to burn helium,
so it cools, starts to fade,
and continues to
decrease in mass.

4 **Brown dwarf**
Gravity continues
to shrink the star so that it
is a fraction of its former
size. It becomes dimmer,
glowing only at infrared
wavelengths.

5 **Black dwarf**
This is the
hypothetical end point
for low-mass stars, as no
star has had enough time
to cool down enough to
become a black dwarf.

Hydrogen fusion
begins in shell
outside core

Helium dumped in
core, which swells

Planetary nebulae often
look spectacular but are
relatively short-lived

White dwarfs can
reach temperatures
exceeding 100,000 K

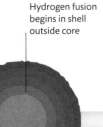

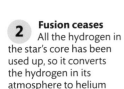

2 **Subgiant stage**
In this phase, the star swells
as it burns helium in its core and
the shell outside the core becomes
hot enough to fuse hydrogen.

3 **Red giant stage**
The star expands
dramatically as the hydrogen
fusion in the shell creates extra
helium to fuel the core.

4 **Planetary nebula**
Eventually, the star throws
out its shells of gas to form a
glowing envelope of clouds
called a planetary nebula.

5 **White dwarf**
As the clouds of
the planetary nebula
dissipate, the old core
remains and becomes
a bright white dwarf.

Supernovae can be
seen across Universe

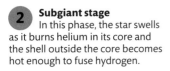

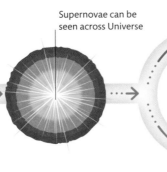

If mass of star is between 1.4
and 3 solar masses, remnant
collapses into neutron star

If remnant of star is over 3 solar
masses, it forms a black hole

2 **Supergiant stage**
Supergiants and
hypergiants are the biggest
stars in the Universe.

3 **Supernova**
When a supergiant has
used up all of its fuel, it collapses
and explodes in a supernova.

4 **Collapsing star**
Depending on its mass,
the remnant collapses into a
neutron star or a black hole.

Red giants

When low- and medium-mass stars use up all the hydrogen in their cores, they reach the end of their long, stable lives on the main sequence. They quickly swell into red giants for the last phases of their lives, becoming much bigger and brighter but glowing a coolish red.

The life cycle of a red giant

Low- and medium-mass stars like the Sun spend 90 percent of their lives on the main sequence of the H-R diagram (see pp.88–89). But eventually, they use up the hydrogen in their cores, which contract and grow hotter until the surrounding shell of hydrogen gets so hot that fusion starts. This makes them swell hugely to become red giant stars approximately 62 million–620 million miles (100 million–1 billion km) in diameter—that is, 100 to 1,000 times the size of the Sun today.

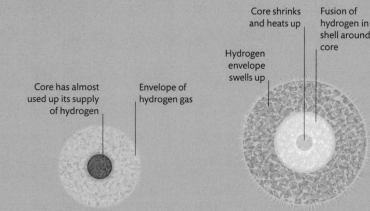

Core shrinks and heats up

Fusion of hydrogen in shell around core

Hydrogen envelope swells up

Core has almost used up its supply of hydrogen

Envelope of hydrogen gas

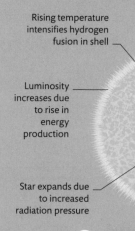

Rising temperature intensifies hydrogen fusion in shell

Luminosity increases due to rise in energy production

Star expands due to increased radiation pressure

1 **Exhausted core**
By now, the star's core has used most of its fuel supply of hydrogen nuclei. There is more hydrogen in the layers outside the core, but it is not hot enough to fuse. The core starts to contract, getting hotter and denser.

2 **Shell ignition**
Hydrogen in the layer over the shrinking core falls inward and heats up. It begins to fuse into helium in a shell surrounding the old core. Driven by this new burst of heat, the star swells quickly.

3 **Bigger and brighter**
Medium-mass stars grow rapidly to become red giants. Hydrogen fusion in the shell around the core dumps helium in the core, which also swells. The boost in energy production makes the star glow brightly.

THE SUN AS A RED GIANT

In about 5 billion years, the Sun will exhaust its hydrogen, begin helium fusion, and turn into a red giant star. As the Sun expands, its outer layers will engulf Mercury, probably Venus, and possibly Earth as well.

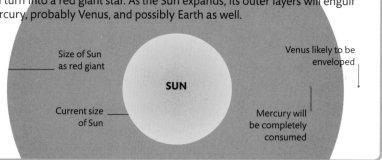

Size of Sun as red giant

Venus likely to be enveloped

SUN

Current size of Sun

Mercury will be completely consumed

WHAT MAKES A RED GIANT RED?

A star's color depends on its surface temperature, which for a typical red giant is about 9,000°F (5,000°C). This puts the brightest light it emits in the orange–red part of the spectrum.

Helium-4 nucleus, also known as alpha particle

Beryllium-8 nucleus created

Gamma ray released

Gamma ray produced

Oxygen-16 nucleus formed

Reversible reaction, as beryllium-8 can decay back into helium-4 nuclei

REACTION

REACTION

REACTION

Helium-4 nucleus

Third helium-4 nucleus joins

Carbon-12 nucleus formed

Helium-4 nucleus

1 **First fusion**
Two helium-4 nuclei fuse, forming a beryllium-8 nucleus. Beryllium-8 is unstable and normally decays in a fraction of a second back into two helium-4 nuclei.

2 **Carbon produced**
In the split second before it decays, a beryllium-8 nucleus may collide with a helium-4 nucleus. This reaction produces a carbon-12 nucleus and energy as a gamma ray.

3 **Oxygen produced**
The carbon-12 nucleus may fuse with another helium-4 nucleus to produce an oxygen-16 nucleus. This reaction also releases a gamma ray.

HELIUM FUSION, OR THE TRIPLE-ALPHA PROCESS

Core becomes denser and hot enough to start helium fusion

Hydrogen fusion ceases in the shell, star shrinks, and luminosity reduces; radiation pressure from core makes shell swell

Fusion of hydrogen in shell reinvigorated

Helium fusion starts in shell

Outer surface heats up again as star shrinks

Carbon core

Luminosity rises as star swells

4 **Helium flash**
Helium fusion (see above) begins suddenly with the "helium flash" in which energy production shoots up 100 billion times. Pressure from the core causes the hydrogen shell to expand, reducing its energy output. This makes the star shrink and become dimmer.

5 **Final burn-out**
Once all the helium in the core has been used, hydrogen and helium fusion continue in two shells around the core. Helium produced in the hydrogen shell fuels the helium shell. Both shells heat up and the star brightens and expands.

Changing temperature and brightness

Once they leave the main sequence, low- and medium-mass stars take a zigzag path across the H-R diagram. Each change in direction across the chart reflects the change in temperature and brightness at different phases in the star's life. The three key stages are: the red-giant branch (RGB); the horizontal branch (HB) that begins with the helium flash (HF); and the final, "asymptotic giant" branch (AGB) when the star has developed a carbon-oxygen core.

Zigzag path across the H-R diagram
The zigzag path of a star with a mass similar to the Sun shows how it first grows cooler even as it gets bigger and brighter, then heats up, before finally cooling again.

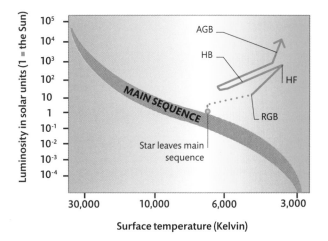

Planetary nebulae

Massive stars explode and low-mass stars fade away, but medium-mass stars become planetary nebulae, which gradually dim and leave behind white dwarfs. They are among the most colorful objects in the Universe.

Knots form in areas more resistant to shockwave

Ultraviolet radiation ionizes shell of gas, which starts to glow

Gaseous tentacles form in envelope

A dying star
In the last stages of its life, a red giant (see pp.110–111) expands at such a speed that gas in its outer layers escapes the star's gravity. This gas is also pushed away by pressure exerted from the star's core.

Ultraviolet radiation from core

3 **Thin shell formed**
The shockwave interacts with the hydrogen and bunches it up into a shell. Gaseous tentacles form in the envelope when expanding hot gas pushes into cooler gas. Ultraviolet light from the brightening central star ionizes the shell and causes it to glow.

Helium shell

Hydrogen shell

Hydrogen envelope blowing off red giant

Fast-moving wind catches up with more slowly moving envelope

White dwarf with exposed core at over 100,000°C (180,000°F)

Carbon core collapsing inward

2 **Radiation emitted**
The star's core contracts further, becoming a bright white dwarf. Intense ultraviolet radiation emitted from the core begins to travel outward, heating the previously ejected hydrogen. The fast stellar wind catches up with the envelope, creating a shockwave.

How a planetary nebula forms

A planetary nebula forms gradually and continually evolves. First, the layers surrounding a red giant's burned-out core are blown off as a fast wind. Then the star's exposed core sends out a brilliant glow of mostly ultraviolet radiation. This is invisible to the naked eye, which is why planetary nebulae do not look as bright as they really are unless false-color imaging is used (see pp.94–95). Despite the name, planetary nebulae have nothing to do with planets; the name arose in the 18th century, when observers thought some of the first to be discovered resembled the disk shape of a planet.

Hydrogen shell layer blows outward as fast wind

1 **Shell blown out**
The old red giant's core collapses, and it expels its burned-out hydrogen shell. The resulting stellar wind blows the shell out in all directions into space, traveling at a speed of approximately 43,500 mph (70,000 kph).

Planetary nebula shapes

There is a huge variety of planetary nebula shapes, but most can be grouped into three types: spherical, elliptical, and bipolar. The variety arises partly because their appearance seems to change when they are viewed from different angles, a phenomenon called the projection effect. But the shape may be also be affected if the central star has a companion, planets, or a magnetic field.

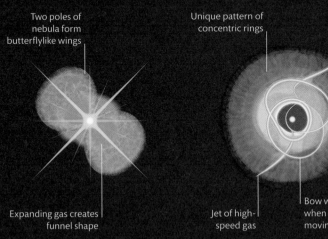

Two poles of nebula form butterflylike wings

Unique pattern of concentric rings

Expanding gas creates funnel shape

Jet of high-speed gas

Bow wave created when gas hits slower-moving material

Butterfly Nebula (bipolar)
This bipolar planetary nebula has two lobes shaped like butterfly wings. Bipolar nebulae such as this are thought to have formed when the central object is a binary system, in which only one star survives.

Cat's Eye Nebula (elliptical)
The bright central part of the beautiful Cat's Eye Nebula is incredibly complex. It is surrounded by a faint halo of rings, blown out like bubbles at intervals of 1,500 years.

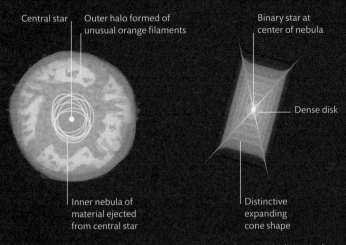

Central star

Outer halo formed of unusual orange filaments

Binary star at center of nebula

Dense disk

Inner nebula of material ejected from central star

Distinctive expanding cone shape

NGC 2392 (spherical)
This nebula reminds some people of a head surrounded by a furry hood. The central structure is due to overlapping bubbles of ejected material.

Red Rectangle Nebula (bipolar)
It is not understood how this uniquely shaped nebula formed. One idea is that gas ejected from its binary star sent out shockwaves after hitting a thick dust ring.

HOW LONG DO PLANETARY NEBULAE LAST FOR?

Planetary nebulae exist for only a short time—tens of thousands of years compared to the star's lifespan of several billion years.

CHEMICAL COMPOSITION

The chemical nature of planetary nebulae is revealed by the spectra of their light (see pp.26–27). A strong red emission line, called a hydrogen alpha line, is caused by a hydrogen electron falling from its third- to second-lowest energy level. This is what often gives planetary nebulae a reddish color. A strong green line reveals a type of ionized oxygen formed only in the low-density setting of a planetary nebula.

Intensity

Hydrogen

Helium

Ionized oxygen

Hydrogen alpha

Wavelength

TYPICAL PLANETARY NEBULA EMISSION SPECTRUM

IN **5 BILLION YEARS,** THE **SUN** WILL BECOME A **FAINT PLANETARY NEBULA**

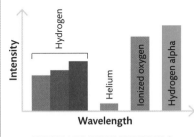

Surface texture consists of hot (bright) and cooler (dark) regions

Pressure exerted by densely packed electrons

Temperature drops rapidly in this zone as heat is being radiated to atmosphere

DEGENERATE CARBON AND OXYGEN INTERIOR

SHELL OF NONDEGENERATE MATERIAL

Gravitational pressure

CRUST

Balanced forces
The pressure exerted by degenerate electrons (see below) balances the force of gravity, preventing the star from collapsing any further. However, this pressure is not enough to keep a white dwarf stable unless its mass is less than 1.4 times the Sun's mass.

Inside a white dwarf

As red giants (see pp.110–111) use up their remaining fuel, they expel their outer layers as planetary nebulae (see pp.112–113), leaving only a tiny, hot core, known as a white dwarf. This remnant slowly cools down and fades. A white dwarf's atmosphere is composed mostly of either hydrogen or helium. The interior, composed mostly of carbon with some oxygen, is thought to crystallize as the white dwarf cools. Because a diamond is crystallized carbon, a white dwarf can be compared to an Earth-sized diamond.

White dwarfs

Sun-sized stars that formed soon after the birth of the Universe end their lives as white dwarfs. They are a little bigger than Earth yet contain about the same amount

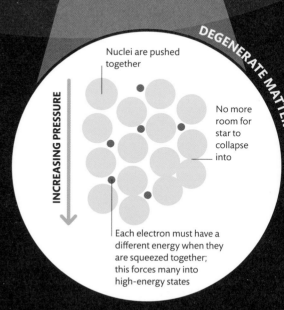

DEGENERATE MATTER

Nuclei are pushed together

INCREASING PRESSURE

No more room for star to collapse into

Each electron must have a different energy when they are squeezed together; this forces many into high-energy states

How degenerate matter forms
Without fusion, there is no energy source to counteract the inward pull of gravity. Gravity crunches the electrons and nuclei much closer than they would be in atoms. This is called a degenerate state. Degenerate matter exerts a

Crust thought to be only
30 miles (50 km) thick

Atmosphere of almost pure
hydrogen or helium

Companion planet

White dwarf star

White dwarfs and planetary destruction

In 2014, scientists working on K2, the second space mission involving the Kepler Space Telescope (see p.187), believed they had observed a white dwarf in the process of destroying its own planetary system. The intense gravitational pull of the white dwarf appeared to be tearing fragments of its companion planet away into orbit around the star, creating a debris disk. A simulation of the process over the course of 120 days, after the planet first begins to feel the significant effects of the star's intense gravitational force, is shown here.

WHO FIRST DETECTED A WHITE DWARF?

Telescope maker Alvan Clark discovered one in 1863. He realized that the slight "wobble" in the star Sirius's orbit was caused by the gravity exerted by a white dwarf companion.

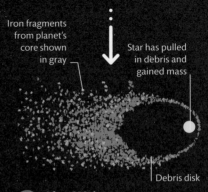

1 After 1 day
The gravitational force of an Earth-sized white dwarf pulls mass from an orbiting planet. The blue line shows a stream of rocky fragments being drawn away from the planet.

Spiral-shaped debris disk starting to form

Rocky fragments broken away from planet

2 After 16 days
More rocky fragments are pulled from the exterior of the planet, which is now rotating faster and faster. A debris disk can be seen forming around the star.

Iron fragments from planet's core shown in gray

Star has pulled in debris and gained mass

Debris disk

3 After 120 days
The planet has completely broken up. The inner part of the debris disk is almost entirely rocky, with iron from the planet's core littered over a wider field. The star has accumulated mass from the destroyed planet.

THE CHANDRASEKHAR LIMIT

Indian-American astrophysicist Subrahmanyan Chandrasekhar discovered that there is a limit to the amount of mass a white dwarf can have and remain stable, supported by its degenerate matter. Beyond that limit, approximately 1.4 times the mass of the Sun, a white dwarf collapses and explodes as a supernova (see pp.118–119), leaving behind either a neutron star or a black hole.

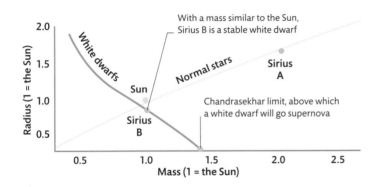

With a mass similar to the Sun, Sirius B is a stable white dwarf

White dwarfs

Normal stars

Sun

Sirius A

Sirius B

Chandrasekhar limit, above which a white dwarf will go supernova

Radius (1 = the Sun)

Mass (1 = the Sun)

Blue supergiant
Blue supergiants, such as Rigel A, are much larger than the Sun but far smaller than red supergiants. These stars have only just come off the main sequence (see pp.88–89) and are incredibly luminous.

Red giant
The brightest star in the constellation Taurus, Aldebaran has a radius 44 times that of the Sun. It is only about 65 light-years from Earth, so it appears as the 14th brightest star in the night sky.

Blue hypergiant
The Pistol Star is one of the brightest stars in the Milky Way, with a luminosity (see p.89) approximately 1.6 million times that of the Sun. It is classified as a blue hypergiant and also thought to be a luminous blue variable star, an as yet not fully understood phase in the life cycles of massive stars.

ALDEBARAN

Many supergiants start off blue, but then expand via yellow to red, getting cooler all the time

RIGEL A

PISTOL STAR

Atmosphere of Antares
Antares is around 700 times larger than the Sun, but an international effort concluding in 2020 showed that its atmosphere, including the lower and upper chromosphere and wind acceleration zones, reaches out 2.5 times farther than that.

Photosphere

Lower chromosphere

Upper chromosphere

Wind acceleration zone

ANTARES

ATMOSPHERE LAYERS

Supergiants

Supergiants are stars of very high mass that have used up the last of their hydrogen fuel and entered the final phases of their lives. At this point in their evolution, they have swollen to enormous sizes.

Comparing sizes
Here, various star sizes are compared to the radius of the Sun. Blue stars tend to be smaller than their red counterparts but are just as bright due to their higher surface temperatures.

The life cycle of a supergiant

Like red giants, supergiants fuse helium when they have depleted their supply of hydrogen before starting to fuse the heavier elements. As they fuse these elements, the stars swell to become supergiants. Supergiants do not live for as long as red giants though, with the largest stars having the shortest lifespans. Supergiants end their lives in spectacular fashion, exploding in supernovae (see pp.118–119).

THE PISTOL STAR RELEASES AS MUCH ENERGY IN 20 SECONDS AS THE SUN DOES IN A YEAR

Stars such as the Pistol Star are rare and exhibit dramatic variations in their brightness

Red supergiant
Antares was estimated at 680 solar radii, but recent measurements suggest it might be much bigger.

Pollux has a radius around nine times greater than the Sun

Bellatrix has a luminosity 9,211 times that of the Sun

Sun is a main sequence star classified in the "G" spectral class (see pp.88–89)

Orange giant
Pollux is an orange giant star in the constellation Gemini. It is approximately 30 times brighter than the Sun and is the closest giant star.

Blue giant
Bellatrix, in the constellation Orion, has a radius 5.75 times that of the Sun. In time, it may evolve into an orange giant.

Yellow dwarf
Although it appears tiny next to the giants and supergiants, our Sun is actually a slightly larger than average star.

HOW BIG CAN A STAR GET?

There does seem to be an upper limit to a star's mass. Collapsing protostars over 150 times more massive than the Sun generate so much energy that they blow themselves apart.

Wolf-Rayet stars

Wolf-Rayet stars are extremely hot and at an advanced stage of evolution. With masses around 10 times that of the Sun, they fuse heavy elements in their cores, which stops them from collapsing under their own immense mass. This generates intense heat and radiation that propels strong stellar winds out at speeds of up to 5.6 million mph (9 million kph). These winds make Wolf-Rayet stars lose mass at a high rate. Many of them have companion stars, and their interacting stellar winds create a distinctive spiral of dust.

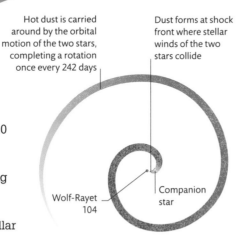

Hot dust is carried around by the orbital motion of the two stars, completing a rotation once every 242 days

Dust forms at shock front where stellar winds of the two stars collide

Wolf-Rayet 104

Companion star

Spiral outflow
Dust formed when the intense stellar winds from Wolf-Rayet 104 and its companion star collide is blown outward and swirled into a spiral by the two stars orbiting each other.

HYPERGIANTS

Hypergiants are the biggest stars in the Universe. It is difficult to determine which is the largest, because they have vague edges and they continually lose mass as their surfaces are blown away by powerful stellar winds. Among the biggest are VY Canis Majoris and UY Scuti, both approximately 1,400 times as big as the Sun.

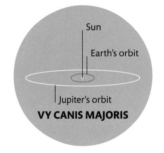

Sun

Earth's orbit

Jupiter's orbit

VY CANIS MAJORIS

Exploding stars

Stars can explode in spectacular phenomena called supernovae. The largest explosions humans have ever seen, supernovae can outshine galaxies for a few days and can even be seen across the Universe.

How stars explode

There are two main categories of supernova. A type II supernova is the natural end point for all high-mass stars that have run out of fuel. The star's core collapses in a quarter of a second, which triggers a colossal shockwave that causes an explosion. Type Ia supernovae happen in binary-star systems when a white dwarf star either collides with or draws too much matter from its neighbor.

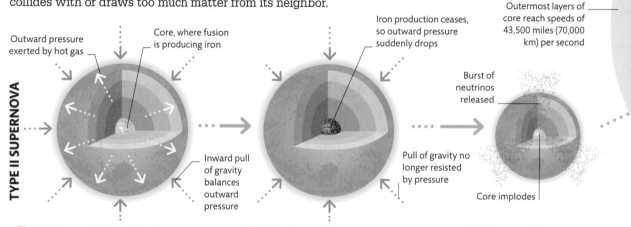

TYPE II SUPERNOVA

Outward pressure exerted by hot gas

Core, where fusion is producing iron

Iron production ceases, so outward pressure suddenly drops

Outermost layers of core reach speeds of 43,500 miles (70,000 km) per second

Burst of neutrinos released

Inward pull of gravity balances outward pressure

Pull of gravity no longer resisted by pressure

Core implodes

1 Red supergiant on the edge
The star is supported by nuclear fusion taking place in its core and the shells around it. The core starts producing iron, but the fuel supply soon runs out.

2 Ready to collapse
When fusion into iron stops, the core collapses because there is not enough outward pressure from hot gas to counteract the inward force of gravity.

3 Core collapses
When the core collapses, it happens in seconds. This sets off a colossal shockwave that makes the outer part of the star explode.

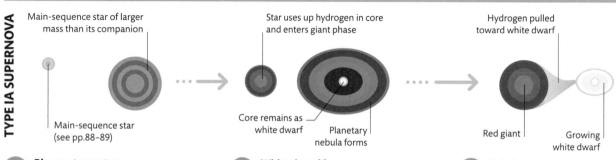

TYPE IA SUPERNOVA

Main-sequence star of larger mass than its companion

Star uses up hydrogen in core and enters giant phase

Hydrogen pulled toward white dwarf

Main-sequence star (see pp.88–89)

Core remains as white dwarf

Planetary nebula forms

Red giant

Growing white dwarf

1 Binary star system
Two stars orbit each other. One of the stars, being of greater mass, approaches the end of its life faster than its companion star.

2 White dwarf forms
The higher-mass star blows off its outer layers, creating a planetary nebula, which exposes a white dwarf. The other star enters the giant phase of its life.

3 Gaining mass
The two stars spiral in closer, and material from the swelling red giant spills onto the white dwarf, increasing its mass toward the maximum it can support.

THE **LAST VISIBLE** **SUPERNOVA** IN THE **MILKY WAY** WAS SEEN IN **1604**

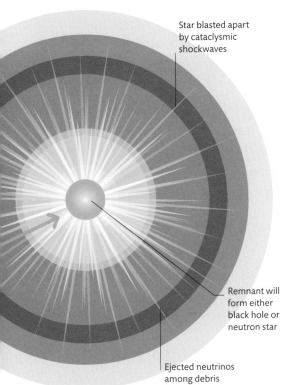

Star blasted apart by cataclysmic shockwaves

Remnant will form either black hole or neutron star

Ejected neutrinos among debris

4 **Star explodes**
The explosion creates an expanding and incredibly bright cloud of hot gas, leaving behind a super-dense remnant core, which may become a black hole or a neutron star depending on the star's mass.

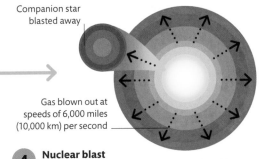

Companion star blasted away

Gas blown out at speeds of 6,000 miles (10,000 km) per second

4 **Nuclear blast**
As more hydrogen accumulates on the white dwarf, it eventually heats up enough for fusion to begin suddenly and explosively. The white dwarf is blown apart and the companion star is ejected away.

Supernovae and heavy elements

Stars are the Universe's chemical forges, creating all the different natural elements. In their cores, stars convert simple elements like hydrogen into heavier elements (see p.91). These include elements, such as carbon and nitrogen, which are needed for life, plus iron, which forms planetary cores. Some of the heavier elements, such as copper and zinc, are made by the force of a supernova, which also scatters them across space.

1 H HYDROGEN	2 He HELIUM	3 Li LITHIUM	4 Be BERYLLIUM	5 B BORON	6 C CARBON
7 N NITROGEN	8 O OXYGEN	9 F FLUORINE	10 Ne NEON	11 Na SODIUM	12 Mg MAGNESIUM
13 Al ALUMINUM	14 Si SILICON	15 P PHOSPHORUS	16 S SULFUR	17 Cl CHLORINE	18 Ar ARGON
19 K POTASSIUM	20 Ca CALCIUM	21 Sc SCANDIUM	22 Ti TITANIUM	23 V VANADIUM	24 Cr CHROMIUM
25 Mn MANGANESE	26 Fe IRON	27 Co COBALT	28 Ni NICKEL	29 Cu COPPER	30 Zn ZINC

Created by stars
This diagram shows the various origins of the 40 lightest elements. Hydrogen and helium formed soon after the Big Bang, but many of the elements were created either by exploding massive stars or by exploding white dwarfs.

KEY
- Big Bang
- Dying low-mass stars
- Cosmic ray fission
- Exploding massive stars
- Exploding white dwarfs

SUPERNOVA SPOTTING

Amateur astronomers can play a part in discovering supernovae by making their own observations of galaxies and by using their computers to examine images of galaxies. Supernovae are named by their year of discovery, prefixed by SN and followed by a letter code.

Pulsars

In the late 1960s, intense, regular radio pulses were detected from deep space. They came from neutron stars emitting powerful pulses as they spin. These stars became known as pulsars, an abbreviation for "pulsating radio star."

HOW DO PULSARS SPIN SO FAST?

The fastest pulsars flash hundreds of pulses a second. These "millisecond" pulsars gain their speed from gases flowing from a companion star, which acts like a jet of water turning a wheel.

Neutron stars

A neutron star is all that remains of a supergiant of over 10 solar masses after it has exploded in a supernova (see pp.118–119). The star collapses so powerfully under its own gravity that it is squeezed into a ball barely 12 miles (20 km) across. In a neutron star, protons and electrons are squeezed together to form a sea of tightly packed neutrons. Neutron stars are the densest objects in the Universe that can be observed directly.

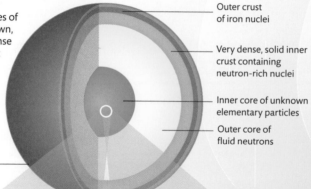

Inside a neutron star
While the outer features of a neutron star are known, the inner core is so dense that scientists have not determined what it comprises. There are several theories, including a traditional view and the hyperon core theory.

Thin atmosphere of carbon plasma

Outer crust of iron nuclei

Very dense, solid inner crust containing neutron-rich nuclei

Inner core of unknown elementary particles

Outer core of fluid neutrons

Neutron stars have hugely powerful magnetic fields, rotating at same speed as star

Star's powerful magnetic field accelerates particles out in a funnel along its two magnetic poles

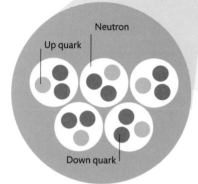

Neutron

Up quark

Down quark

Traditional theory
This theory suggests that the inner core may consist of tightly packed neutrons containing three quarks—two "down" and one "up" quark.

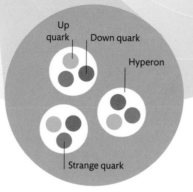

Up quark

Down quark

Hyperon

Strange quark

Hyperon core theory
This theory indicates that under extreme pressure, a down quark could change into a "strange" quark, creating a subatomic particle called a hyperon.

Celestial lighthouse
Neutron stars that emit directed beams of radiation are known as pulsars. They are characterized by their strong magnetic fields and fast rotation. Over time, their rotation speed slows down as they lose energy.

13,000 BILLION LB (6 BILLION TONNES)

THE MASS OF A **TEASPOON** OF **NEUTRON STAR** MATERIAL

COSMIC COLLISION

Two neutron stars can orbit each other, like binary stars. If they move close enough, they may spiral to their own destruction. These collisions, called kilonovas, emit bursts of gamma rays and may be the source of much of the Universe's gold, platinum, and other heavy elements. In 2017, gravitational waves reached Earth from a kilonova that occurred around 130 million years ago.

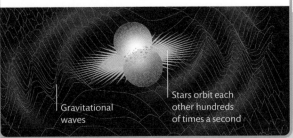

Gravitational waves

Stars orbit each other hundreds of times a second

Speed of rotation comes from rapid collapse of star

NEUTRON STAR

Neutron star's gravity is so strong that its solid surface, which is around a million times stronger than steel, is pulled into a smooth sphere

How a pulsar works

Most of the roughly 3,000 neutron stars that have been found are pulsars. Without the powerful beam of radio waves that pulsars emit, neutron stars are so tiny that they would otherwise be hard to see. Pulsars are like cosmic lighthouses, sending out pairs of radio beams that sweep across the Universe as they rotate, typically once every 0.25–2 seconds. Radio telescopes on Earth only spot pulsars at the moment that their beams sweep across Earth.

PULSAR "ON"

As a pulsar rotates, its two radiation beams continually sweep through space. At the instant shown here, one of the radiation beams points at Earth. This can be detected on Earth as a brief radio signal.

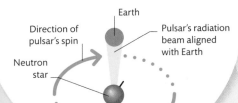

Earth

Direction of pulsar's spin

Pulsar's radiation beam aligned with Earth

Neutron star

PULSAR "OFF"

At the moment shown here, neither of the radiation beams emanating from the pulsar points at Earth, so from the perspective of an observer on Earth, the pulsar is "off."

Radiation beam not aligned with Earth

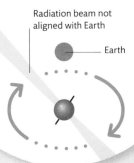

Earth

SUPERMASSIVE BLACK HOLES ARE THOUGHT TO LIE AT THE CENTER OF MOST LARGE GALAXIES

How a black hole forms

Once a massive star has exploded in a supernova and its core collapses beyond a certain point, it becomes a stellar black hole. Matter that is pulled toward the black hole by gravity can form a spinning disk, releasing radiation that can be detected by astronomers. Supermassive black holes are thought to form after stars collide or through the merging of many smaller black holes.

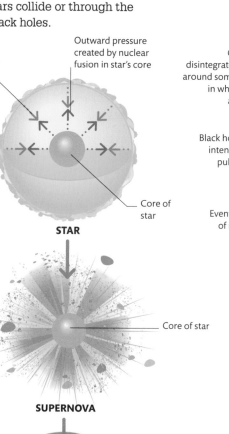

1 A stable star
Nuclear reactions in the core of a star create energy and outward pressure. When these are in balance with the force of gravity pulling inward, the star remains stable. But when the fuel runs out, gravity takes over.

Inward gravitational force

Outward pressure created by nuclear fusion in star's core

Core of star

STAR

2 Spectacular end
When a massive star exhausts its fuel, the nuclear reactions cease and the star dies. Unable to resist the crushing force of its own gravity, the star collapses. A supernova explosion then blasts the star's outer layers into space.

Core of star

SUPERNOVA

3 Core collapses
If the core that remains after the supernova is more than three times the mass of the Sun, nothing can stop it from collapsing. It will keep on shrinking until it reaches a point of infinite density called a singularity.

Gravitational force

Singularity

CORE OF DYING STAR

MATTER JOINING ACCRETION DISK

ACCRETION DISK

Gas, dust, and disintegrated stars spiral around some black holes in what is called an accretion disk

Black hole forms area of intense gravity, which pulls matter inward like a whirlpool

Event horizon is point of no return for any matter or light that crosses it from outside

EVENT HORIZON

GRAVITY WELL

INCREASING INTENSITY OF GRAVITY

Concealed in center of black hole is an infinitely small and dense singularity, where matter has been squeezed

Black holes

Black holes are regions in space where gravity is so strong that it sucks everything in, including light. A black hole can form when the core of a massive star turns to iron and implodes under gravity.

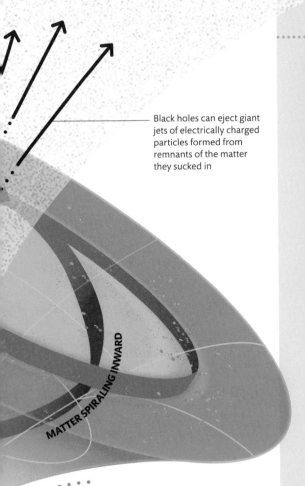

Black holes can eject giant jets of electrically charged particles formed from remnants of the matter they sucked in

MATTER SPIRALING INWARD

4 **A black hole forms**
By now, the density of the singularity is so great that it distorts space-time surrounding it so that not even light can escape. A black hole can be pictured as an infinitely deep hole called a gravity well.

WHAT IS A WORMHOLE?

It is a theoretical tunnel through the curved fabric of space-time (see pp.154–155). Something could enter a wormhole at one point in space-time and emerge in another.

Types of black hole

There are two main types of black hole: stellar and supermassive. Stellar black holes form when an old supergiant star collapses in a supernova. From the number of giant stars in the Milky Way, scientists estimate there could be up to a billion such black holes in this galaxy alone. Supermassive black holes are far larger than stellar black holes and are thought to have masses up to billions of times that of the Sun. There is also evidence for a third, midsized type that is intermediate in mass between stellar and supermassive black holes.

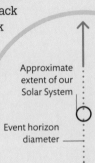

Approximate extent of our Solar System

Event horizon diameter

Diameter of event horizon of Holm 15a, the most massive black hole known

Contrasting sizes
While stellar black holes are relatively small, the Holm 15a supermassive black hole, discovered in 2019, is thought to be 40 billion times the mass of the Sun.

STELLAR
Event horizon diameter: 20–200 miles (30–300 km)
Mass: 5–100 Suns

SUPERMASSIVE
Event horizon diameter: thousands of light-years
Mass: billions of Suns

SPAGHETTIFICATION

Approaching a black hole's event horizon, the gravitational pull increases so significantly that objects dragged toward it are stretched into long, spaghettilike strands. An astronaut would be torn apart, legs first, by this "spaghettification" process. Time would run at different speeds for his head and feet.

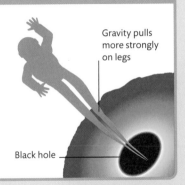

Gravity pulls more strongly on legs

Black hole

GALAXIES AND THE UNIVERSE

The Milky Way

Our galaxy, the Milky Way, is a medium-sized spiral galaxy. It is only one of the two trillion galaxies in the Universe—groups of stars, gas, and dust held together by gravitational attraction.

The structure of the Milky Way

The Milky Way is a typical spiral galaxy. It has an elongated bulge, or nucleus, at its center, with a supermassive black hole at its very core (see pp.128–129). Two major spiral arms—the Scutum-Centaurus Arm and the Perseus Arm—extend from each end of the central bar, and there are also several minor arms. The arms form a thin disk 100,000–120,000 light-years across. There is also a spherical halo of stars about 170,000–200,000 light-years in diameter.

Regions between arms contain lower density of gas, dust, and stars

THOUSANDS OF LIGHT-YEARS FROM CENTER

50   40   30   20   10

SAGITTARIUS ARM

FAR 3KPC

PERSEUS ARM

ORION SPUR

OUTER ARM

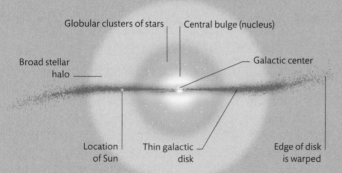

Globular clusters of stars · Central bulge (nucleus)

Broad stellar halo

Galactic center

Location of Sun

Thin galactic disk

Edge of disk is warped

Side view of the Milky Way

Precise measurements of the positions of Cepheid variable stars (see p.98), shown in green, have shown that our galaxy is warped at its edges. This warping may have been the result of a past collision with another, smaller galaxy.

Spiral arms contain relatively high density of gas, dust, and stars

HOW MANY STARS ARE IN THE MILKY WAY?

Most stars are too dim to be easily observed, but the Milky Way is thought to contain 100–400 billion stars.

Anatomy of the Milky Way

Our galaxy's core is densely packed with old, yellow stars. The stars in the spiral arms are younger and bluer. Dark lanes of dust crisscross the arms, some fringed with glowing red nebulae of ionized gas. The oldest stars are outside the disk in globular star clusters that form part of a broad, sparsely populated stellar halo.

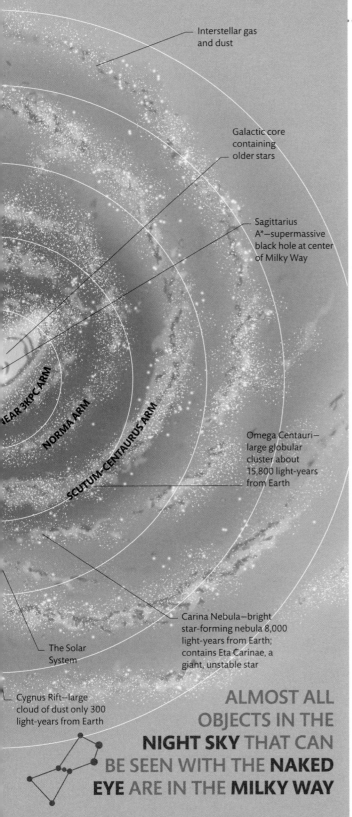

Interstellar gas and dust

Galactic core containing older stars

Sagittarius A*—supermassive black hole at center of Milky Way

NEAR 3KPC ARM

NORMA ARM

SCUTUM-CENTAURUS ARM

Omega Centauri—large globular cluster about 15,800 light-years from Earth

Carina Nebula—bright star-forming nebula 8,000 light-years from Earth; contains Eta Carinae, a giant, unstable star

The Solar System

Cygnus Rift—large cloud of dust only 300 light-years from Earth

ALMOST ALL OBJECTS IN THE NIGHT SKY THAT CAN BE SEEN WITH THE NAKED EYE ARE IN THE MILKY WAY

Our local neighborhood

The Sun lies about 26,000 light-years from the galactic center, on the edge of the Orion Spur. We are in a bubble of hot, ionized (electrically charged) hydrogen gas surrounded by clouds of cooler dust and molecular hydrogen gas (in which each hydrogen molecule is in the form of two linked atoms) alive with star-forming nebulae. Neighboring bubbles are outlined by loops of glowing interstellar dust.

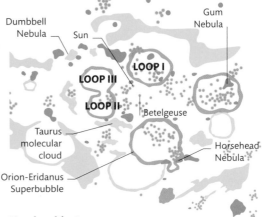

Dumbbell Nebula

Sun

Gum Nebula

LOOP I

LOOP III

LOOP II

Betelgeuse

Taurus molecular cloud

Horsehead Nebula

Orion-Eridanus Superbubble

Nearby objects

This map of the Milky Way's local neighborhood shows part of the Orion Arm. The Sun is toward the center; hydrogen gas clouds are shown in yellow, gas and dust clouds in red, and star clusters and giant stars are blue.

THE MILKY WAY IN THE SKY

The Milky Way appears as a bright, whitish, hazy band, densely populated with stars, running across the night sky. When we look at the band, we are looking into the depths of our galaxy's disk.

Cygnus Rift dust cloud obscures part of Milky Way

MILKY WAY

MILKY WAY FROM THE NORTHERN HEMISPHERE

The center of the Milky Way

The nucleus of our galaxy takes the form of a central bulge that extends for about 800 light-years. Densely packed with stars, it contains the supermassive black hole Sagittarius A* at its center.

The galactic center

The nucleus of our galaxy is obscured at visible light wavelengths by dust. However, it can be studied using other wavelengths, such as infrared and radio waves, which can penetrate the dust. A strong source of radio waves known as Sagittarius A lies at the center of our galaxy. It consists of Sagittarius A* (often abbreviated to Sgr A*), a supermassive black hole; Sagittarius A East, a supernova remnant; and Sagittarius A West, a collection of gas and dust falling into Sgr A*. Shorter-wavelength X-rays and gamma rays are emitted from the center, indicating intense activity, with dust and gas being accelerated to extremely high speeds.

The Milky Way's hub
Most of the stars in the central region of our galaxy are old red giants, although there are also a few younger stars orbiting close to Sagittarius A*, which were possibly formed in the disk of gas there.

HOW DO WE KNOW WHERE THE CENTER OF THE MILKY WAY IS?

All of the objects in the Milky Way appear to revolve around the supermassive black hole Sagittarius A*, so it must be the center of our galaxy.

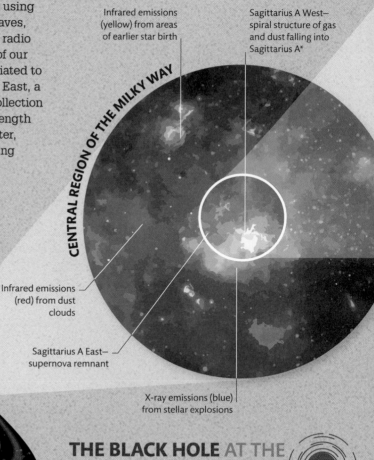

CENTRAL REGION OF THE MILKY WAY

Infrared emissions (yellow) from areas of earlier star birth

Sagittarius A West— spiral structure of gas and dust falling into Sagittarius A*

Infrared emissions (red) from dust clouds

Sagittarius A East— supernova remnant

X-ray emissions (blue) from stellar explosions

THE MILKY WAY

Spiral arms

Galactic nucleus, densely packed with old stars

Direction of rotation of disk around nucleus

THE BLACK HOLE AT THE CENTER OF THE MILKY WAY HAS A MASS EQUAL TO ABOUT 4.3 MILLION SUNS

The heart of the Milky Way
At the very center of our galaxy is an area of strong radio emissions where material is being pulled in and torn apart by a supermassive black hole—Sagittarius A*. The black hole cannot be seen directly, but astronomers have confirmed its existence and measured its mass by tracking stars orbiting close to it.

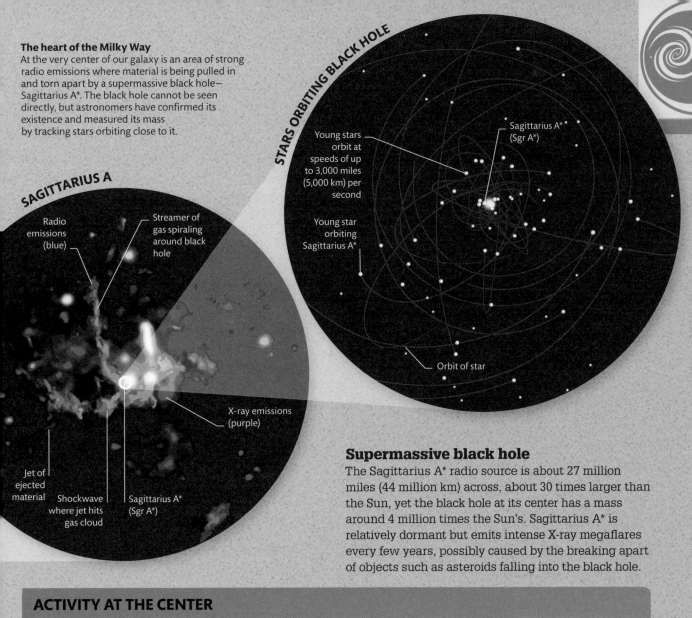

STARS ORBITING BLACK HOLE

Young stars orbit at speeds of up to 3,000 miles (5,000 km) per second

Sagittarius A* (Sgr A*)

Young star orbiting Sagittarius A*

Orbit of star

SAGITTARIUS A

Radio emissions (blue)

Streamer of gas spiraling around black hole

X-ray emissions (purple)

Jet of ejected material

Shockwave where jet hits gas cloud

Sagittarius A* (Sgr A*)

Supermassive black hole

The Sagittarius A* radio source is about 27 million miles (44 million km) across, about 30 times larger than the Sun, yet the black hole at its center has a mass around 4 million times the Sun's. Sagittarius A* is relatively dormant but emits intense X-ray megaflares every few years, possibly caused by the breaking apart of objects such as asteroids falling into the black hole.

ACTIVITY AT THE CENTER

Giant lobes of gas extend for thousands of light-years above and below the galactic center, funneled by streams of X-ray-emitting gas. These bubbles were discovered by the Fermi spacecraft, which detected the gamma rays also emitted by the gas. Gamma rays are the form of electromagnetic radiation that carry the most energy (see p.153).

Radiation emissions
The radiation emissions from the galactic center are due to movement of material—possibly particle jets or gas from an earlier burst of star formation—away from the supermassive black hole Sgr A*.

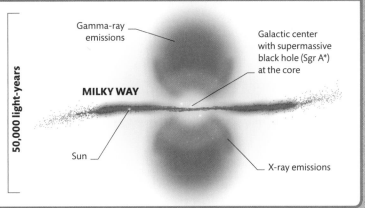

Gamma-ray emissions

Galactic center with supermassive black hole (Sgr A*) at the core

MILKY WAY

50,000 light-years

Sun

X-ray emissions

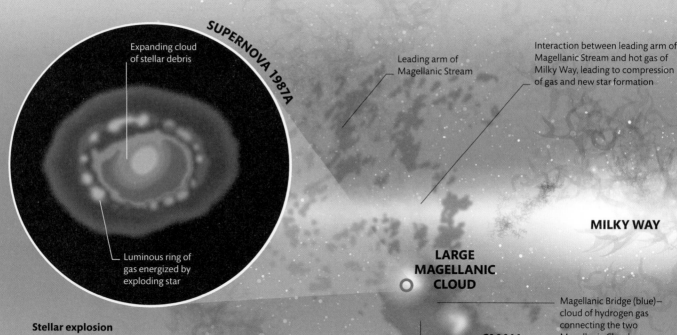

SUPERNOVA 1987A

Expanding cloud
of stellar debris

Leading arm of
Magellanic Stream

Interaction between leading arm of
Magellanic Stream and hot gas of
Milky Way, leading to compression
of gas and new star formation

Luminous ring of
gas energized by
exploding star

MILKY WAY

**LARGE
MAGELLANIC
CLOUD**

Magellanic Bridge (blue)—
cloud of hydrogen gas
connecting the two
Magellanic Clouds

**SMALL
MAGELLANIC
CLOUD**

Hydrogen gas
from SMC pulled
out by stronger
gravity of LMC

Magellanic Stream (red)—flow of
high-velocity hydrogen gas linking
Magellanic Clouds with Milky Way

Stellar explosion
In 1987, a star in the LMC went
supernova, blazing with the power of
100 million Suns, the brightest explosion
seen from Earth in the last 400 years.

Large Magellanic Cloud

The Large Magellanic Cloud (LMC) is a dwarf spiral
galaxy (see pp.140–141) with a prominent central bar
and spiral arm. The gravitational pull of the Milky Way
makes it a site of vigorous star formation. Like the
Milky Way, the LMC contains globular and open star
clusters, planetary nebulae, and clouds of gas and dust.

The Magellanic Clouds

Named after Ferdinand Magellan, the
Portuguese explorer who observed them as
he sailed south of the equator in 1519, the
Magellanic Clouds are a spectacular feature of
the night sky in the southern hemisphere. Lying
in the constellations Dorado and Tucana near
the south celestial pole, these irregular clouds of
stars are small galaxies in their own right and
two of the Milky Way's closest neighbors.

WHO DISCOVERED THE MAGELLANIC CLOUDS?

The clouds have been known
since ancient times by
indigenous peoples of the
southern hemisphere. The first
written references to them are
by Arab scholars in about the
9th century CE.

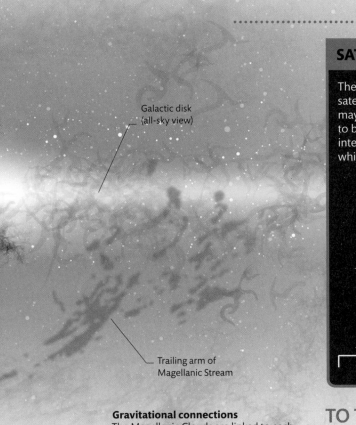

Galactic disk
(all-sky view)

Trailing arm of
Magellanic Stream

Gravitational connections
The Magellanic Clouds are linked to each
other by a cloud of hydrogen gas and to
the Milky Way by a stream of fast-moving
hydrogen gas. These structures are a result
of the gravitational interaction between
the clouds and the Milky Way.

SATELLITES OR PASSERSBY?

The Magellanic Clouds are generally considered to be
satellite galaxies orbiting the Milky Way. However, they
may be independent bodies just passing by. They seem
to be moving too fast to be long-term satellites, but this
interpretation depends on the mass of the Milky Way,
which is uncertain.

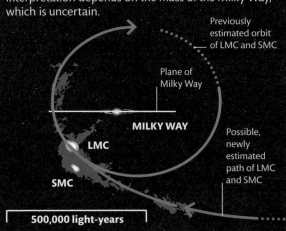

Previously
estimated orbit
of LMC and SMC

Plane of
Milky Way

MILKY WAY

LMC

Possible,
newly
estimated
path of LMC
and SMC

SMC

500,000 light-years

TO THE **NAKED EYE**, THE
MAGELLANIC CLOUDS
APPEAR AS **FAINT, IRREGULAR
PATCHES** IN THE SOUTHERN SKY

Small Magellanic Cloud

An irregular dwarf galaxy, the
Small Magellanic Cloud (SMC)
is one of the most distant objects
visible to the naked eye. It has the
remnant of a central bar, which
suggests that it may have been
a barred spiral before it was
disrupted by the gravitational
influence of the Milky Way. There
is also gravitational interaction
between the two Magellanic
Clouds: the SMC orbits around
the LMC, and they share a
common cloud of hydrogen gas—
the Magellanic Bridge—that
is a region of star formation.

	THE MAGELLANIC CLOUDS COMPARED		
	The SMC is more distant, smaller, less massive, and has fewer stars than the LMC. Both are dwarf galaxies, but the SMC is an irregular galaxy, whereas the LMC is a dwarf spiral.		
		LMC	**SMC**
	DISTANCE FROM EARTH	163,000 light-years	200,000 light-years
	DIAMETER	14,000 light-years	7,000 light-years
	MASS	80 billion Suns	40 billion Suns
	NUMBER OF STARS	10–40 billion	Several hundred million

WHEN WAS THE ANDROMEDA GALAXY DISCOVERED?

The galaxy was first identified as a "nebulous smear" in the night sky by the Persian astronomer Al-Sufi in around 964 CE.

The Andromeda Galaxy

Andromeda is the closest large galaxy to the Milky Way and the brightest and largest of the Local Group (see pp.134–135). It is a barred spiral, like the Milky Way, and studying Andromeda has helped us understand the nature of our own galaxy.

THE **ANDROMEDA GALAXY** IS ON COURSE TO **COLLIDE** WITH **THE MILKY WAY** IN ABOUT **5 BILLION YEARS**

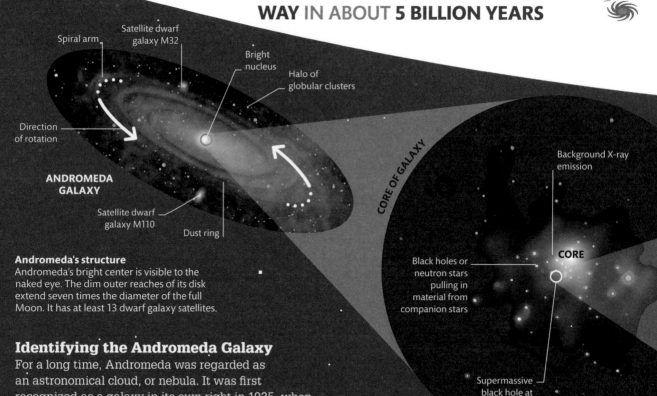

Spiral arm

Satellite dwarf galaxy M32

Bright nucleus

Halo of globular clusters

Direction of rotation

ANDROMEDA GALAXY

Satellite dwarf galaxy M110

Dust ring

CORE OF GALAXY

Background X-ray emission

CORE

Black holes or neutron stars pulling in material from companion stars

Supermassive black hole at center of galaxy

Andromeda's structure
Andromeda's bright center is visible to the naked eye. The dim outer reaches of its disk extend seven times the diameter of the full Moon. It has at least 13 dwarf galaxy satellites.

Identifying the Andromeda Galaxy

For a long time, Andromeda was regarded as an astronomical cloud, or nebula. It was first recognized as a galaxy in its own right in 1925, when Edwin Hubble calculated the distance to its Cepheid variable stars (see pp.98–99) and proved that they lay outside the Milky Way. Located about 2.5 million light-years from Earth, Andromeda is visible to the naked eye, but it is difficult to make out its structure because it lies almost edge-on to our view. However, infrared observations have revealed that it is a barred spiral galaxy with at least one huge ring of dust.

Galactic core
X-ray observations of Andromeda reveal 26 stellar black holes (see p.123) or neutron stars in its central bulge. Their intense gravitational fields are pulling material in from companion stars in binary star systems, releasing high-energy radiation. A super-massive black hole lies at the very centre of the galaxy.

The structure of the galaxy

Distinct populations of stars can be seen in the Andromeda Galaxy: young blue stars in the spiral arms of the disk (and around the central black hole); and old red stars in the central bulge. The same pattern of star distribution is also found in our own galaxy. The Andromeda Galaxy has prominent, dark dust lanes, where most star formation is taking place, but these dust lanes are more circular than spiral in shape. A relatively small dust ring in the inner part of the galaxy may have resulted from an encounter with M32, a neighboring dwarf galaxy in the Local Group, at least 200 million years ago.

COMPARING THE ANDROMEDA GALAXY AND THE MILKY WAY

Andromeda is twice the size and has twice the number of stars, but its overall mass might be the same or even lower compared to the Milky Way.

Andromeda Galaxy
- **Galaxy type:** Barred spiral galaxy
- **Diameter:** 220,000 light-years (excluding halo)
- **Mass:** 1,000 billion Suns
- **Number of stars:** 1,000 billion
- **Number of globular clusters:** 460

Andromeda's spiral arms are fragmented and may be transitioning to a more ringlike structure.

Milky Way
- **Galaxy type:** Barred spiral galaxy
- **Diameter:** 100,000–120,000 light-years (excluding halo)
- **Mass:** 850–1,500 billion Suns
- **Number of stars:** 100–400 billion
- **Number of globular clusters:** 150–158

The Milky Way has a well-defined spiral structure for both the stars and the dust lanes in its disk.

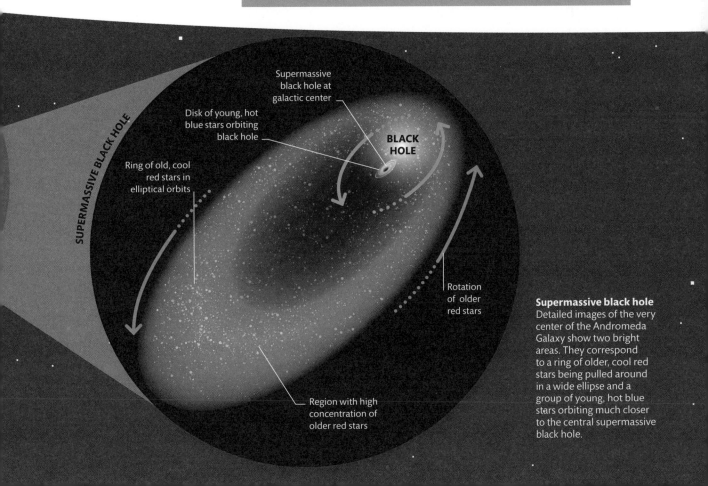

SUPERMASSIVE BLACK HOLE

Supermassive black hole at galactic center

Disk of young, hot blue stars orbiting black hole

BLACK HOLE

Ring of old, cool red stars in elliptical orbits

Rotation of older red stars

Region with high concentration of older red stars

Supermassive black hole
Detailed images of the very center of the Andromeda Galaxy show two bright areas. They correspond to a ring of older, cool red stars being pulled around in a wide ellipse and a group of young, hot blue stars orbiting much closer to the central supermassive black hole.

4 MILLION LIGHT-YEARS

3 MILLION LIGHT-YEARS

2 MILLION LIGHT-YEARS

1 MILLION LIGHT-YEARS

Sextans B

Sextans A

Leo A

NGC 3109

Antlia Dwarf

Leo I

Leo II

Canes Dwarf

Ursa Major I

Sextans Dwarf

Ursa Major II

Boötes Dwarf

Ursa Minor Dwarf

Draco Dwarf

Large Magellanic Cloud

Milky Way

Small Magellanic Cloud

Sagittarius Dwarf

Carina Dwarf

Sculptor Galaxy

Fornax Dwarf

Andromeda I

Barnard's Galaxy

Phoenix Dwarf

Aquarius Dwarf

SagDIG

IC 1613

Tucana Dwarf

Cetus Dwarf

WLM (Wolf–Lundmark–
Melotte) Galaxy

HOW MANY GALAXIES ARE IN THE LOCAL GROUP?

More than 50 galaxies have been identified, but the total number is likely to remain unknown, as some will be forever hidden behind the Milky Way.

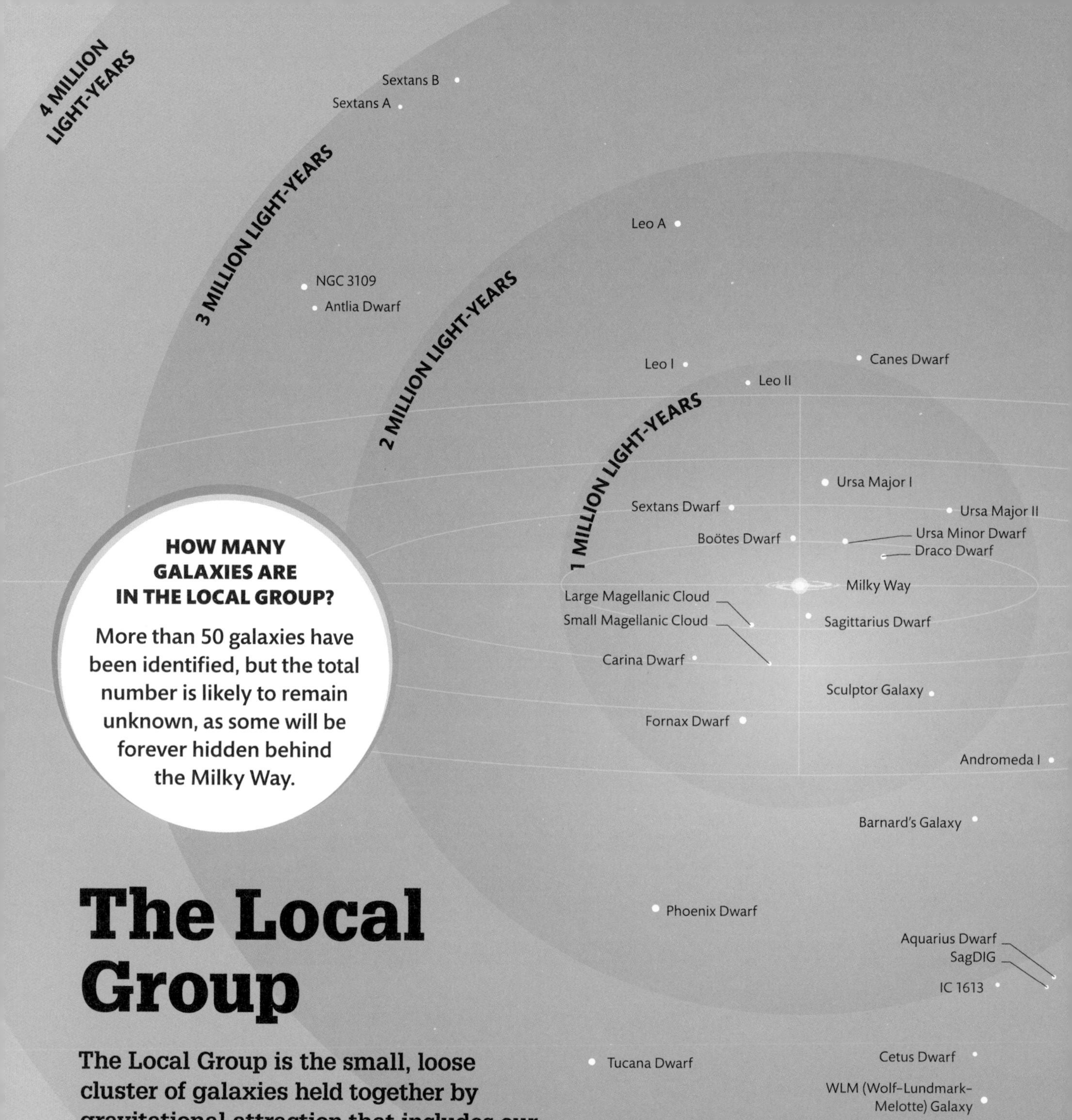

The Local Group

The Local Group is the small, loose cluster of galaxies held together by gravitational attraction that includes our Milky Way Galaxy (see pp.126–129) and the Andromeda Galaxy (see pp.132–133), its largest members. Most of the others are dwarf galaxies (see pp.140–141).

The Local Group galaxies
Most Local Group galaxies are satellites of the Milky Way or Andromeda. The distant Antlia-Sextans Group forms a subgroup, and there are also several small, independent galaxies. This view is centred on the Milky Way, but all the galaxies in the group actually orbit a centre of mass between the Milky Way and Andromeda galaxies.

The evolution of the Local Group

The Local Group is relatively young, so most of its gas is still contained within its galaxies, feeding star formation. The Milky Way's largest neighbors—the Magellanic Clouds (see pp.130–131)—are being pulled in by the gravity of their parent. Similarly, the Milky Way and Andromeda galaxies are moving closer together and will ultimately merge. The Local Group itself may one day merge with the nearest neighboring galaxy cluster, the much larger Virgo Cluster (see pp.146–147).

IC 10

NGC 185

NGC 147

M110

Andromeda Galaxy

M32

Andromeda II

Andromeda III

Triangulum Galaxy

Pisces Dwarf

Pegasus Dwarf

THE ESTIMATED **MASS** OF THE **LOCAL GROUP** IS **2 TRILLION TIMES** THE **MASS OF THE SUN**

The Triangulum Galaxy

Lying about 2.7 million light-years away, the Triangulum Galaxy is one of the most distant objects visible to the naked eye. It is the third largest member of the Local Group, with a diameter of about 60,000 light-years. Triangulum had a close encounter with the Andromeda Galaxy about 2–4 billion years ago, triggering star formation in Andromeda's disk.

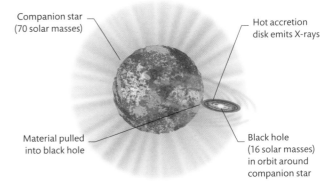

Companion star (70 solar masses)

Hot accretion disk emits X-rays

Material pulled into black hole

Black hole (16 solar masses) in orbit around companion star

Stellar black hole
The Triangulum Galaxy contains an unusual binary star system consisting of a black hole with about 16 times the mass of the Sun orbiting a much more massive star. X-rays are emitted as material from the star is pulled into the black hole.

BARNARD'S GALAXY

Barnard's Galaxy contains many areas of intense star formation, such as the Bubble, Ring, Hubble V, and Hubble X nebulae. Lying about 1.6 million light-years away, Barnard's Galaxy was one of the first systems outside our galaxy to have its distance calculated, from observations of its Cepheid variable stars (see pp.98–99).

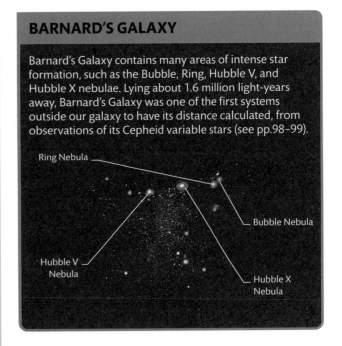

Ring Nebula

Bubble Nebula

Hubble V Nebula

Hubble X Nebula

Spiral galaxy structure

Spiral galaxies have a flattened disk rich in stars, gas, and dust. This material is concentrated into a number of arms that spiral around a central bulge, which is densely packed with stars and sometimes elongated into a bar. The spiral arms are bright with young blue stars, whereas older red and yellow stars dominate in the central bulge and in an extensive halo, which includes globular star clusters.

ABOUT TWO-THIRDS OF ALL OBSERVED GALAXIES ARE SPIRALS

Stars in spiral galaxies

In a typical spiral galaxy, most of the stars are situated in the flat galactic disk and in the spherical bulge of the nucleus around the central black hole. Some stars are also found in a broad, spherical halo, usually in compact globular star clusters.

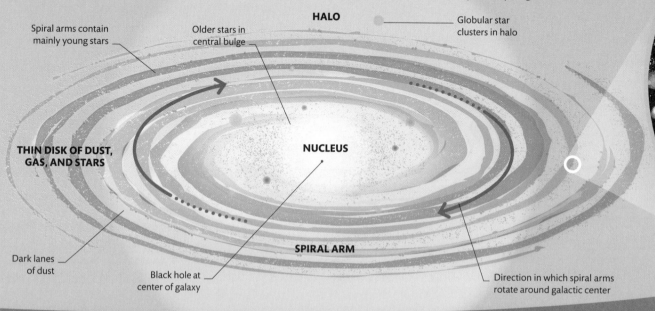

Spiral arms contain mainly young stars

Older stars in central bulge

HALO

Globular star clusters in halo

THIN DISK OF DUST, GAS, AND STARS

NUCLEUS

Dark lanes of dust

Black hole at center of galaxy

SPIRAL ARM

Direction in which spiral arms rotate around galactic center

Spiral galaxies

Spiral galaxies are some of the most spectacular objects in the Universe. Their appearance depends on density variations within their disks, which determines the number of spiral arms, how tightly they are wound, and how distinct they are.

Spiral arms

A galaxy is not a solid structure but a fluid collection of stars, gas, dust, and other objects, all rotating about the galaxy's center. Spiral arms originate as waves of high density in this material, which rotate more slowly than the material itself. Stars and gas enter a density wave in much the same way as cars enter a traffic jam, bunching up and moving through it and out to the other side. This bunching up triggers the creation of bright new stars that we see as the spiral arms.

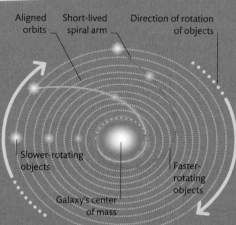

Aligned orbits

Short-lived spiral arm

Direction of rotation of objects

Slower-rotating objects

Galaxy's center of mass

Faster-rotating objects

Idealized galaxy

In an ideal galaxy, with objects moving at the same speed in aligned orbits, outer objects take longer to complete their orbits than those nearer the center. Although spiral patterns develop, the spirals are soon so tightly wound that they become indistinct.

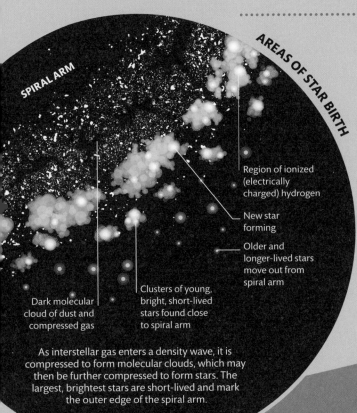

SPIRAL ARM

AREAS OF STAR BIRTH

Region of ionized (electrically charged) hydrogen

New star forming

Older and longer-lived stars move out from spiral arm

Dark molecular cloud of dust and compressed gas

Clusters of young, bright, short-lived stars found close to spiral arm

As interstellar gas enters a density wave, it is compressed to form molecular clouds, which may then be further compressed to form stars. The largest, brightest stars are short-lived and mark the outer edge of the spiral arm.

Activity in the spiral arms

The spiral arms are slow-moving density waves in the galactic disk, which give rise to areas of intense star formation as gas is compressed on entering the area of higher density. The brightest newborn stars emit lots of ultraviolet light, which ionizes hydrogen in the gas (splits hydrogen molecules into electrically charged particles) and causes it to glow. These bright stars and glowing gas are what give definition to the spiral arms.

WHICH IS THE LARGEST SPIRAL GALAXY?

In 2019, the Hubble Space Telescope imaged one of the largest known spiral galaxies, UGC 2885. Located about 232 million light-years away, it is about 2.5 times wider than the Milky Way and contains 10 times as many stars.

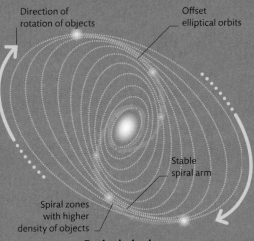

Direction of rotation of objects

Offset elliptical orbits

Stable spiral arm

Spiral zones with higher density of objects

Real spiral galaxy
In a real galaxy, outer objects still take longer to complete their orbits than inner ones, but their orbits are elliptical and are at slightly different angles. Over time, this leads to the objects bunching together in some places, producing the effect of stable spiral arms.

STAR ORBITS

Stars within the disk bob up and down while following elliptical orbits around the center, roughly in the plane of the galaxy. Stars in the central bulge have short orbits at random angles, leading to a spherical distribution a few hundred light-years across. Similarly, stars in the halo orbit at all angles, but they plunge through the disk on long orbits that can take them thousands of light-years above and below the galactic plane.

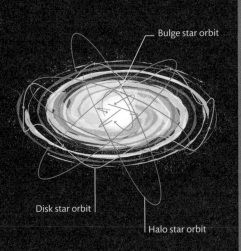

Bulge star orbit

Disk star orbit

Halo star orbit

Elliptical galaxies

Elliptical galaxies are smooth balls of stars with little structure. They span a vast range of sizes, and their shape varies from oval to spherical. The biggest are far larger than any spiral galaxies. Lenticular galaxies share some features of ellipticals but also have certain similarities to spiral galaxies.

WHICH IS THE LARGEST KNOWN GALAXY?

The elliptical galaxy IC 1101 is the largest known galaxy of any type. It contains about 100 trillion stars and has a halo up to 4 million light-years across.

Oval-shaped halo containing old yellow and red stars and many globular clusters

Galaxy contains little dust or gas

Orbits tilted at any angle and with a large range of eccentricity

Anatomy of an elliptical galaxy
M86 is a typical elliptical galaxy similar in size to the Milky Way but containing about 300 times as many globular clusters. It does not have a well-defined nucleus, and the star density decreases smoothly with distance from the center.

Orbits in elliptical galaxies
Elliptical galaxies have little interstellar dust and gas to interact with the stars and keep them flattened into a single plane, so the orbits of the stars are chaotic, inclined at any angle and varying in shape from circular to eccentric ellipses.

Elliptical galaxies

These galaxies vary enormously in size, from about a tenth the size of the Milky Way to supergiants tens of times wider than our galaxy. Ellipticals contain mostly older yellow and red stars with low mass. They have little interstellar gas or dust, and very little star formation occurs within them, probably because almost all of their gas and dust has already been turned into stars. A giant elliptical galaxy is often the central and brightest member of a galaxy cluster, but dwarf ellipticals are relatively dim and difficult to discover.

Giant elliptical galaxies
Ellipticals are some of the largest galaxies known. Compared to the Milky Way (a typical barred spiral galaxy), M87 is about 10 times wider; IC 1101, one of the largest galaxies currently known, is about 40 times wider. Both of these ellipticals contain many trillions of stars, compared to the hundreds of billions in the Milky Way.

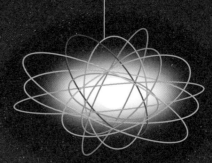

MILKY WAY
Barred spiral galaxy
170,000–200,000 light-years across; 100–400 billion stars

M87
Giant elliptical galaxy
1 million light-years across
Several trillion stars

IC 1101
Supergiant elliptical galaxy
4 million light-years across
About 100 trillion stars

Lenticular galaxies

Lenticular galaxies have a similar appearance to ellipticals, especially when seen side-on, but like spiral galaxies, they have a disk of gas and dust that flattens them into a lens shape—hence the name lenticular, which means lenslike. Some lenticulars may be spiral galaxies that have lost most, but not all, of their gas and dust. Like elliptical galaxies, lenticulars contain older stars and show little sign of new star formation.

DWARF ELLIPTICALS **ARE** DIM **AND** DIFFICULT TO OBSERVE, **BUT THEY ARE** PROBABLY **THE** MOST COMMON **TYPE OF** GALAXY

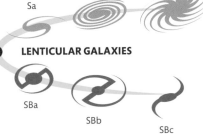

Large, spherical nucleus of older stars

Circular dust lanes

Disk of gas, dust, and older stars

More chaotic elliptical orbits in nucleus

Almost circular orbits in disk

Anatomy of a lenticular galaxy
NGC 2787 is a lenticular galaxy that has a little more structure than most lenticulars, with concentric rings of dust in its disk. Like most lenticulars, NGC 2787 has a larger nucleus than a spiral galaxy of similar size.

Orbits in lenticular galaxies
Stars typically follow well-ordered, almost circular paths in the disk of a lenticular galaxy. However, in the large central bulge, the stars' orbits are more varied and eccentric and are inclined at any angle.

GALAXY CLASSIFICATION

Galaxies are commonly classified according to their shape, and a system still widely used today is the one devised by Edwin Hubble in 1926. He grouped galaxies into three main types according to their shape as seen from Earth: elliptical, lenticular, and spiral. These are commonly represented in a "tuning fork" diagram. The Hubble system is not intended to explain galaxy evolution, and we now recognize a fourth type: irregular galaxies, which do not have a distinct, regular shape (see p.141).

Hubble's galaxy classification
Ellipticals are numbered from E0 (circular) to E7 (highly elliptical). All lenticulars are classed as SO. Spirals are split into classic (S) and barred (SB) types.

CLASSIC SPIRAL GALAXIES

Sa
Sb
Sc

E0
E3
E5
E7
S0

ELLIPTICAL GALAXIES

LENTICULAR GALAXIES

SBa
SBb
SBc

BARRED SPIRAL GALAXIES

Dwarf galaxies

Most of the approximately 2 trillion galaxies in the observable Universe are much smaller than the Milky Way. Some of these dwarf galaxies have definite shapes, such as spirals, but many are irregular.

Galaxy sizes
Dwarf galaxies are typically 10 times smaller than the Milky Way and contain about a hundred-fold fewer stars (less than a few billion).

Milky Way
170,000–200,000 light-years across

Cigar Galaxy
40,000 light-years across

NGC 4449
20,000 light-years across

Large Magellanic Cloud
14,000 light-years across

NGC 1569
8,000 light-years across

Small Magellanic Cloud
7,000 light-years across

Zwicky 18
3,000 light-years across

Features of dwarf galaxies

Most dwarf galaxies are held by the gravitational fields of larger galaxies, orbiting around them like planets around a star. However, some dwarf galaxies are moving independently of any larger body, and others are found in extreme isolation in the gaps between galaxy clusters. Dwarf galaxies are thought to have formed early in the life of the Universe, producing some of the very first stars before merging with neighbors to form larger galaxies (see pp.168–169). There are about 60 dwarf galaxies near the Milky Way; the biggest are the Large and Small Magellanic Clouds (see pp.130–131).

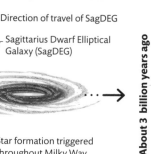

About 6 billion years ago

Direction of travel of SagDEG

Sagittarius Dwarf Elliptical Galaxy (SagDEG)

Star formation triggered throughout Milky Way

FIRST PASS THROUGH MILKY WAY

About 3 billion years ago

Evolution of Milky Way's spiral arms influenced by SagDEG

Stream of stars stripped from SagDEG

SETTLES INTO ORBIT AROUND MILKY WAY

Sagittarius Dwarf interaction
The Sagittarius Dwarf Elliptical Galaxy has crashed through the Milky Way's disk at least three times, triggering star formation each time and slightly warping the disk of the Milky Way. The Sun was formed at about the time of the first encounter.

WHAT IS OUR NEAREST NEIGHBORING GALAXY?

The Canis Major Dwarf Galaxy is only 25,000 light-years away, so it is closer to us than we are to the center of our galaxy.

ABOUT A **QUARTER** OF ALL **KNOWN GALAXIES** ARE THOUGHT TO BE **IRREGULAR**

Irregular galaxies

Many dwarf galaxies are classified as irregular, although infrared observations have revealed that some, such as the Magellanic Clouds, have spiral or barred spiral structures. Because their mass is small, dwarf galaxies are easily pulled around and apart by the more powerful gravitational fields of larger, more massive neighbors, disrupting their original structures. Full-size galaxies can also be irregular. Many of these larger irregular galaxies show evidence of collisions with other galaxies, with distorted remnants of spiral structures or bright areas of star formation—starbursts.

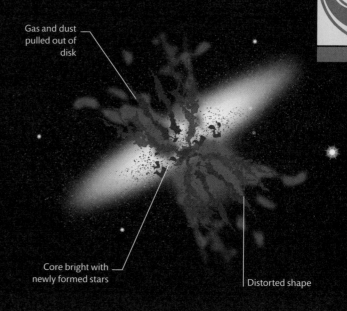

Gas and dust pulled out of disk

Core bright with newly formed stars

Distorted shape

Starburst galaxy
An irregular starburst galaxy, the Cigar Galaxy is being distorted by the gravity of its larger neighbor, M81 (not visible in this image), triggering a high rate of star formation in its core.

About 2 billion years ago

SagDEG sends ripples across Milky Way's disk

SECOND PASS THROUGH MILKY WAY

About 1 billion years ago

Stream of stars from SagDEG encircles Milky Way

THIRD PASS THROUGH MILKY WAY

Present day

SagDEG about 70,000 light-years from Earth

ORBITING MILKY WAY

TYPES OF DWARF GALAXY

Dwarf galaxies are classified according to their shape, features, and composition. As well as the spiral, elliptical, and irregular types found in full-size galaxies, dwarf galaxies also include several unique types, such as compact dwarfs.

	Dwarf elliptical galaxies	Smaller and fainter than ordinary ellipticals; possibly remnants of low-mass spirals or young galaxies		**Dwarf spiral galaxies**	Dwarf spirals are relatively rare; most are located outside galaxy clusters, far from gravitational interactions
	Dwarf spheroidal galaxies	Small, low-luminosity galaxies similar to globular clusters but differentiated from them by having more dark matter		**Compact dwarf galaxies**	Blue compact dwarfs contain young, hot, massive stars; ultra-compact dwarfs are even smaller and tightly packed with stars
	Dwarf irregular galaxies	Small galaxies with no distinct shape; thought to be similar to the earliest galaxies formed in the Universe		**Magellanic spiral galaxies**	Dwarf galaxies with only one spiral arm, like the Large Magellanic Cloud; intermediate between dwarf spiral and irregular galaxies

Active galaxies

Some galaxies are unusually powerful, emitting more energy than their stars alone could produce. When viewed in some parts of the electromagnetic spectrum (see p.153), they can be a thousand times brighter than the Milky Way. These galaxies have an active nucleus that releases a huge amount of energy as matter falls into the central black hole.

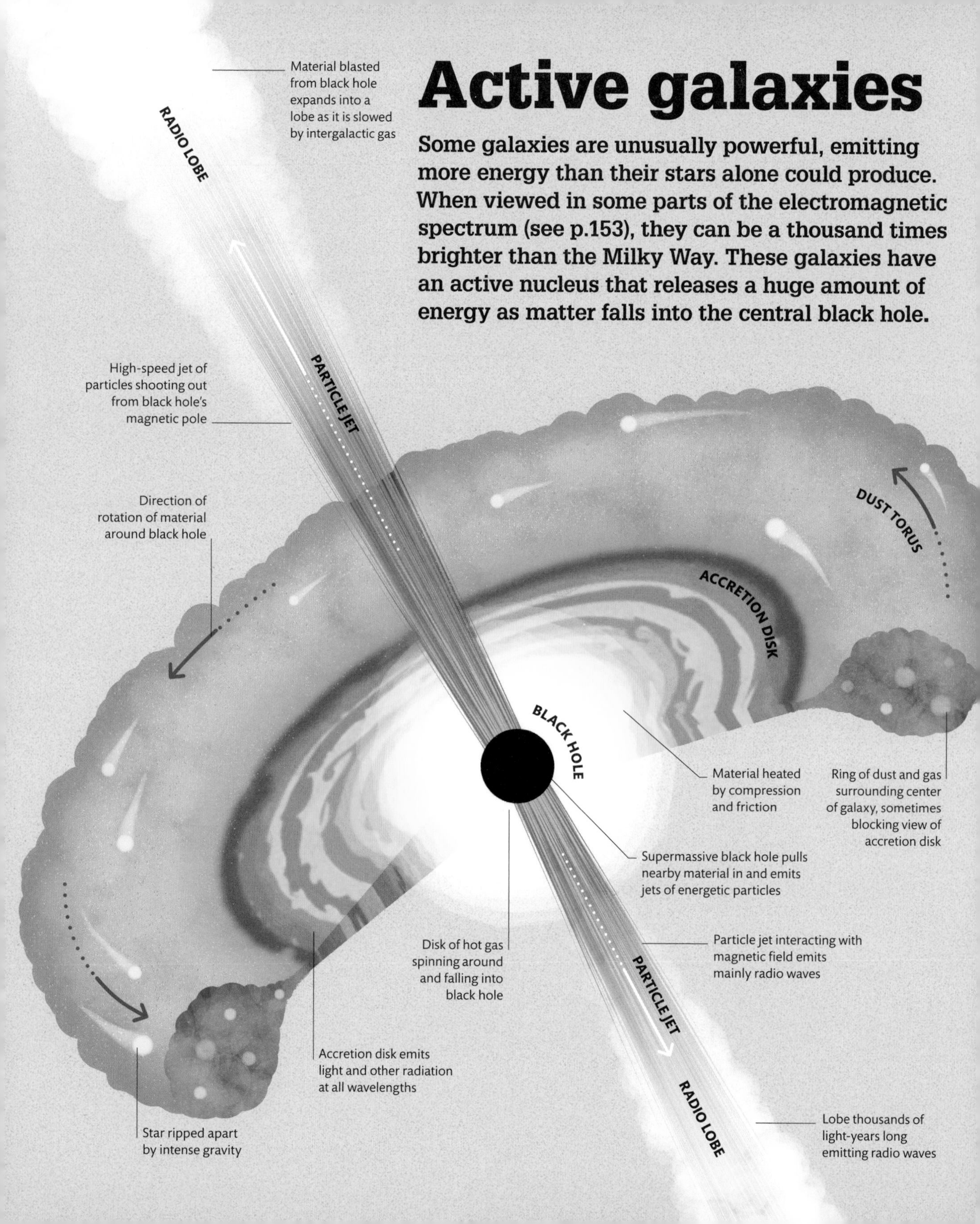

RADIO LOBE

Material blasted from black hole expands into a lobe as it is slowed by intergalactic gas

PARTICLE JET

High-speed jet of particles shooting out from black hole's magnetic pole

Direction of rotation of material around black hole

DUST TORUS

ACCRETION DISK

BLACK HOLE

Material heated by compression and friction

Ring of dust and gas surrounding center of galaxy, sometimes blocking view of accretion disk

Supermassive black hole pulls nearby material in and emits jets of energetic particles

Disk of hot gas spinning around and falling into black hole

Particle jet interacting with magnetic field emits mainly radio waves

PARTICLE JET

Accretion disk emits light and other radiation at all wavelengths

RADIO LOBE

Star ripped apart by intense gravity

Lobe thousands of light-years long emitting radio waves

IS THE MILKY WAY ACTIVE?

Currently, our galaxy is dormant, but the presence of lobes of gamma rays above and below the galactic disk indicates that it may have been active a few million years ago.

Extreme energy

In active galaxies, the central supermassive black hole is consuming nearby matter, which forms a swirling disk that is compressed and heated as it is pulled in and torn apart. Up to a third of the mass pulled into the black hole is turned into energy, making active galaxies the most powerful long-lived objects in the sky. Most active galaxies are very distant from our galaxy, although a few are nearby, and all galaxies have the potential to become active.

Anatomy of an active galaxy

An accretion disk of heated material and a ring (torus) of dust surround the central black hole. Some active galaxies also have huge lobes of radio wave emissions, fed by jets of charged particles from the black hole's magnetic field.

HANNY'S VOORWERP

Hanny's Voorwerp is an unusual object discovered in 2007. Glowing with ionized (electrically charged) oxygen, it was lit up by radiation from a quasar in nearby galaxy IC 2497. The quasar is no longer active, but gas is still streaming from the galaxy, triggering star formation in the ionized cloud.

IC 2497

Star-forming region

Streaming gas

Cloud of gas ejected from IC 2497 by earlier galaxy collision or near-miss

Types of active galaxy

Radio galaxies, Seyfert galaxies, quasars, and blazars are all types of active galaxy emitting X-rays and other forms of high-energy radiation. The type depends on the energy of the activity in the galaxy's nucleus, the mass of the galaxy, and its orientation to Earth. Seyfert galaxies and quasars (quasi-stellar objects) have similar orientations, but Seyferts emit far less energy than quasars, which are among the most powerful and luminous celestial objects known.

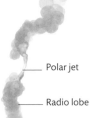

Polar jet

Radio lobe

Accretion disk

Dust ring

RADIO GALAXY NGC 383

Radio galaxy
In a radio galaxy, the central region of the nucleus is hidden by the edge-on dust ring, and observers on Earth see only the polar jets and radio lobes.

QUASAR PG 0052+251

Quasar
In quasars, the dust ring is tilted toward Earth, allowing us to see the brilliant light of the accretion disk, which outshines the light of the surrounding galaxy.

Nucleus emits polar jet

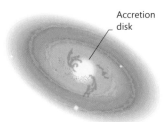

Accretion disk

BLAZAR MAKARIAN 421

Blazar
A blazar is aligned so that observers on Earth look straight down the polar jet into the nucleus. The galaxy is hidden by the brilliant light, but the radio lobes can sometimes be detected.

SEYFERT GALAXY M106

Seyfert galaxy
A Seyfert galaxy has the accretion disk exposed to our view, as it is in a quasar, but the activity in the nucleus is weaker, which allows us to see the surrounding galaxy more clearly.

LIGHT FROM THE MOST **DISTANT QUASARS** HAS TAKEN **MORE THAN 12 BILLION YEARS** TO REACH US

Galaxy collisions

Packed together in clusters, galaxies are large relative to the distances between them, so close encounters and even collisions are common. Collisions can stimulate new star formation and also play a key role in galaxy evolution.

Galaxy interactions

When two galaxies come close, the outcome depends on how large they are and how close they approach. Their interaction may be minor, leading to slight distortion of their shapes, but a major interaction or collision can have dramatic effects, leading to bursts of new star formation or even tearing one or both galaxies apart. A collision can pull material out of a galaxy. It may also propel it into the central black hole, creating an active nucleus (see pp.142–143).

WHAT HAPPENS TO PLANETS WHEN GALAXIES COLLIDE?

When galaxies collide, the gravitational disruption may shift some planets in their orbits or even throw them out into interstellar space, but a collision between planets is very unlikely.

Arm of Whirlpool Galaxy pulled out by gravity of dwarf galaxy now connects the two galaxies

Spiral arms containing young, hot blue stars

NGC 5195 DWARF GALAXY

WHIRLPOOL GALAXY

Shape of dwarf galaxy distorted by collision

Active nucleus emits radiation from matter being pulled into central black hole

Bright pink areas of active star formation

Nucleus glows brightly due to high density of stars and high rate of star formation

Clouds of gas and dust disrupted by collision, leading to new star formation

Whirlpool Galaxy collision
The spiral Whirlpool Galaxy collided with a much smaller dwarf galaxy, NGC 5195, about 300 million years ago, distorting its spiral structure and leading to bursts of star formation. The Whirlpool Galaxy has an active nucleus, possibly as a result of the collision.

Galaxy evolution

Collisions are key to the transformation of one type of galaxy to another. Colliding galaxies may distort each other beyond recognition, or the larger one may engulf the smaller one. A spiral galaxy may be stripped of all its gas and dust, ending star formation and transforming it into an elliptical. Multiple collisions produce giant ellipticals, with their stars orbiting at random angles and any structure of their original constituents lost.

The merger model

According to one theory of galaxy evolution, galaxies undergo a series of mergers and collisions as their interstellar gas is consumed by star formation. The mergers form giant elliptical galaxies that eventually dominate the central areas of galaxy clusters.

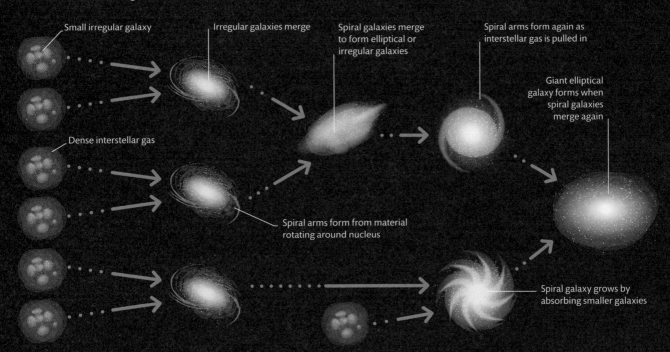

Small irregular galaxy

Dense interstellar gas

Irregular galaxies merge

Spiral galaxies merge to form elliptical or irregular galaxies

Spiral arms form again as interstellar gas is pulled in

Giant elliptical galaxy forms when spiral galaxies merge again

Spiral arms form from material rotating around nucleus

Spiral galaxy grows by absorbing smaller galaxies

THE **MERGER** OF TWO **LARGE GALAXIES** CAN GENERATE **NEW STARS** TOTALING **THOUSANDS OF TIMES THE SUN'S MASS** EVERY YEAR

SIMULATING GALAXY COLLISIONS

Collisions between galaxies happen over millions of years, so it is impossible to observe the whole process. However, computer models using simplified, virtual galaxies can be used to simulate a collision to see what the fate of the galaxies might be. Here, a simulation shows how the structure of two galaxies is disrupted as they collide and merge over a period of a billion years.

0 MILLION YEARS → **500 MILLION YEARS** → **750 MILLION YEARS** → **1 BILLION YEARS**

Galaxy clusters and superclusters

Although some galaxies exist in isolation, most are found in crowds. Their immense gravity pulls them together into small groups, large clusters, and even larger superclusters, some of the largest structures in the Universe.

Superclusters

Galaxy clusters (see below) are themselves grouped into superclusters. Superclusters lie along filaments and sheets between largely empty voids in space (see pp.150–151). There are millions of superclusters in the Universe. The variations that have been detected in the cosmic microwave background radiation (see pp.164–165)—the "echo" of the Big Bang—suggest that these large-scale concentrations of matter date from very early in the life of the Universe. Tiny differences in temperature and matter density during this time gave rise to the first dwarf galaxies, which interacted with their neighbors to grow into galaxy groups, clusters, and superclusters.

Laniakea Supercluster
Our local supercluster, to which the Milky Way and the Local Group belong, is the Laniakea Supercluster. Several nearby superclusters, including the Virgo Supercluster, are now considered to be part of this larger structure.

HOW BIG IS THE LARGEST SUPERCLUSTER?

The Caelum Supercluster, the largest detected, is about 910 million light-years across and contains about half a million galaxies.

50–1,000
THE NUMBER OF GALAXIES IN A TYPICAL GALAXY CLUSTER

THE MISSING MASS

The mass of the stars in a cluster's galaxies does not provide enough gravitational attraction to hold the cluster together. Intergalactic gas provides much more of a cluster's mass, and even more exists as dark matter. Gravitational lensing (see pp.148–149) can help map a cluster's dark matter, which is distributed more broadly than the visible matter we see as galaxies.

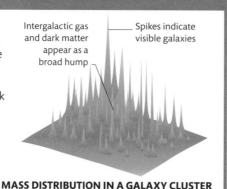

Intergalactic gas and dark matter appear as a broad hump

Spikes indicate visible galaxies

MASS DISTRIBUTION IN A GALAXY CLUSTER

Groups and clusters

Clusters may be relatively sparse, like our Local Group (see pp.134–135), or more densely packed, like the nearby Virgo Cluster. But regardless of how many galaxies they contain, clusters all tend to occupy a similar volume of space, a few million light-years across. The most populous clusters have a dense, spherical distribution of giant elliptical galaxies at their center.

How clusters evolve
From an initial mixture of all galaxy types, collisions and mergers lead to ever larger galaxies and a predominance of elliptical galaxies (see pp.138–139). As a cluster forms, the gas in the cluster becomes hot. The hot gas surrounds and fills the space between the individual galaxies in the cluster.

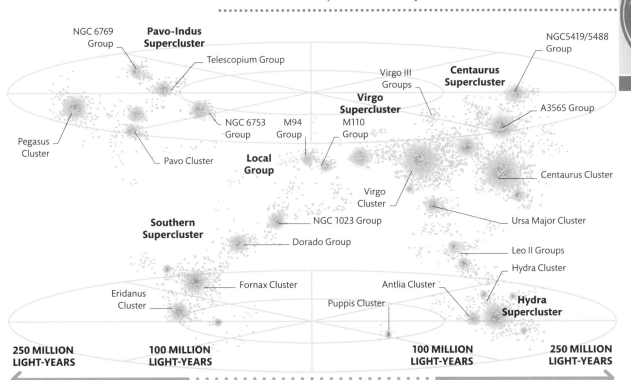

NGC 6769 Group

Pavo-Indus Supercluster

Telescopium Group

NGC5419/5488 Group

Virgo III Groups

Centaurus Supercluster

Virgo Supercluster

A3565 Group

Pegasus Cluster

NGC 6753 Group

M94 Group

M110 Group

Pavo Cluster

Local Group

Centaurus Cluster

Virgo Cluster

Ursa Major Cluster

Southern Supercluster

NGC 1023 Group

Dorado Group

Leo II Groups

Hydra Cluster

Eridanus Cluster

Fornax Cluster

Antlia Cluster

Puppis Cluster

Hydra Supercluster

250 MILLION LIGHT-YEARS

100 MILLION LIGHT-YEARS

100 MILLION LIGHT-YEARS

250 MILLION LIGHT-YEARS

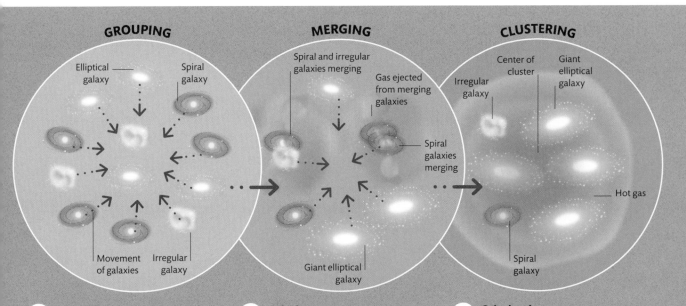

GROUPING

Elliptical galaxy

Spiral galaxy

Movement of galaxies

Irregular galaxy

MERGING

Spiral and irregular galaxies merging

Gas ejected from merging galaxies

Spiral galaxies merging

Giant elliptical galaxy

CLUSTERING

Center of cluster

Giant elliptical galaxy

Irregular galaxy

Hot gas

Spiral galaxy

1 **Loose collection of galaxies**
Clusters begin as a loose, uneven distribution of small galaxies of all types, gravitationally attracted to each other and toward their common center of mass. Many of these galaxies will collide and merge.

2 **Galaxies merge**
When galaxies collide or merge, cold interstellar gas is energized and ejected from the galaxies and a cloud of hot gas, mainly hydrogen, accumulates between the members of the cluster.

3 **Galaxies cluster**
Eventually, giant elliptical galaxies, with old stars and little gas, are densely packed around the cluster's center, cocooned in a spherical cloud of intergalactic gas many times more massive than the galaxies' stars.

Dark matter

Dark matter is matter that is always invisible because, unlike ordinary matter (also called baryonic matter), it does not interact with electromagnetic radiation (see pp.152–153).

How do we know dark matter exists?

Dark matter cannot be observed directly. Instead, its presence has only been detected because of its gravitational influence on visible matter. The idea of dark matter was first put forward in the 1930s to explain why a cluster of galaxies stayed together even though the gravity of the visible galaxies was not strong enough. Then, in the 1970s, the outer regions of galaxies were found to be moving far too fast, indicating invisible matter beyond was pulling them. Now scientists use a technique called gravitational lensing to detect large dark objects and X-rays to detect rises in temperature in interstellar clouds as they are compressed by dark matter.

WHY DO SCIENTISTS BURY THEIR DARK MATTER DETECTORS DEEP UNDERGROUND?

Detectors are buried up to 1.2 miles (2 km) underground to shield them from cosmic rays reaching Earth from space.

GALAXY CLUSTER

Light bent toward observer by cluster acting as lens

How much is missing?
Scientists think just 5 percent of the Universe's mass is ordinary matter. The "missing" portion is dark matter and the even more mysterious dark energy (see p.170).

DARK MATTER 26.8%

ORDINARY MATTER 4.9%

DARK ENERGY 68.3%

Galaxy cluster containing large amount of dark matter acts as a gravitational lens

Contour lines join points of equal dark-matter concentration

Gravitational lensing
When light from distant galaxies is bent by gravity as it passes close to an intervening galaxy cluster, their images are distorted, an effect called gravitational lensing. Dark matter increases the effect, revealing its presence to astronomers and enabling them to map it.

TELESCOPE ON EARTH

Mapping dark matter
By using software to analyze the distorted image of the distant galaxy, astronomers can create a map of the distribution of dark matter in the intervening galaxy cluster.

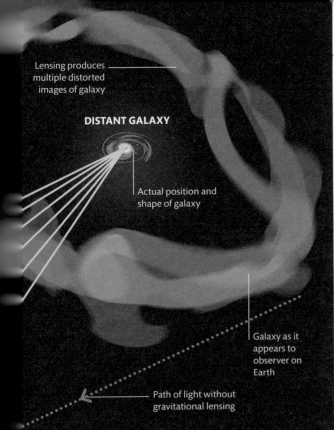

Lensing produces multiple distorted images of galaxy

DISTANT GALAXY

Actual position and shape of galaxy

Galaxy as it appears to observer on Earth

Path of light without gravitational lensing

Types of dark matter

Scientists have envisaged two general candidates for dark matter. MACHOs are large objects made from ordinary baryonic matter that happen not to emit much light. However, these probably account for just a few percent of all dark matter. Scientists now think that we may be entirely immersed in a sea of WIMPs—nonbaryonic subatomic particles that barely interact with light at all.

TYPES OF DARK MATTER		
MACHOs	**WIMPs**	
Some dark matter might consist of dense objects that emit so little light they can be detected only by studying gravitational lensing. Collectively called MACHOs (MAssive Compact Halo Objects), they include black holes and brown dwarfs. However, MACHOs cannot account for all of dark matter's mass.	Dark matter might also include Weakly Interacting Massive Particles (WIMPs), particles that are so called because they can pass through ordinary matter with little or no effect.	
	Hot	**Cold**
	This theoretical form of dark matter consists of particles traveling close to the speed of light.	Most dark matter, such as WIMPs, is thought to be cold—a relatively slow-moving form of matter.

-460°F (-273°C)
THE TEMPERATURE TO WHICH SOME DARK MATTER DETECTORS HAVE TO BE COOLED

Looking for dark matter

If dark matter is subatomic particles that interact only with gravity, then detecting them is difficult. As well as studying the effects of dark matter in space, scientists are also trying to find cold dark matter particles called axions directly by using icy tanks of liquid inert elements buried far below Earth's surface.

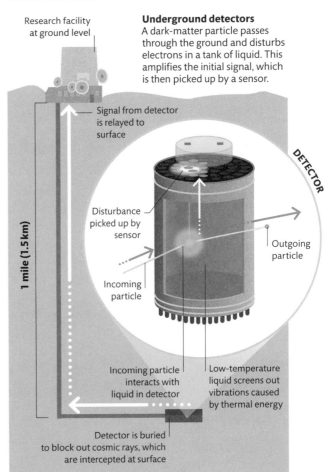

Research facility at ground level

Underground detectors
A dark-matter particle passes through the ground and disturbs electrons in a tank of liquid. This amplifies the initial signal, which is then picked up by a sensor.

Signal from detector is relayed to surface

Disturbance picked up by sensor

DETECTOR

Outgoing particle

Incoming particle

1 mile (1.5 km)

Incoming particle interacts with liquid in detector

Low-temperature liquid screens out vibrations caused by thermal energy

Detector is buried to block out cosmic rays, which are intercepted at surface

Mapping the Universe

In the last 50 years, cosmologists have mapped the Universe in ever more detail. Powerful sky surveys have enabled them to plot similarities and differences across space and to identify vast structures.

THE **LARGEST KNOWN VOID** IN THE COSMIC WEB IS **2 BILLION LIGHT-YEARS** ACROSS

The cosmological principle

According to the cosmological principle, on the largest scales, the Universe is the same everywhere—matter is spread evenly and obeys the same laws. It is both homogeneous (the same in wherever you are) and isotropic (the same whichever direction you look). If this is true, it means that what astronomers see in one area of the Universe is likely to be the same everywhere, and they can simply scale up. But recent observations have thrown doubt on whether it really is homogeneous.

Filaments and voids
The Universe seems to be arranged like a vast cobweb, with all the stars and galaxies concentrated in threadlike filaments and sheetlike walls. In between are dark, empty voids.

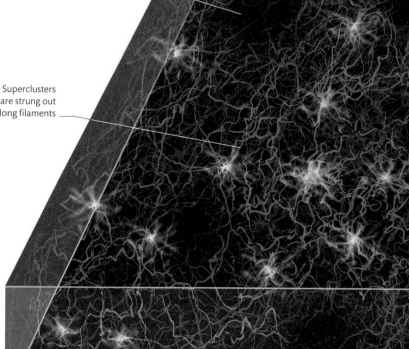

Clusters of galaxies are concentrated at nodes, where filaments meet

Threadlike filaments consist mainly of hot hydrogen gas

Voids are vast and almost spherical

Superclusters are strung out along filaments

4 million light-years

Milky Way Galaxy

Andromeda Galaxy

150 million light-years

Galaxies can be seen to be grouped into clusters

1.5 billion light-years

No structure can be detected in distribution of galaxies

Scale and structure
In theory, there are no structures at the largest scales and the differences that create structures emerge only on smaller scales.

> ## WHAT IS THE BIGGEST STRUCTURE IN THE UNIVERSE?
>
> The largest structure of galaxies found so far is the Sloan Great Wall, nearly 1.5 billion light-years long and about 1 billion light-years from Earth.

SKY SURVEYS

Much of our knowledge of the Universe's large-scale structure is based on 3D maps from surveys of samples of the observable Universe (see pp.160–161). In 2020, the Sloan Digital Sky Survey (SDSS) produced the largest, most detailed map so far, charting the history of the Universe back through 11 billion years.

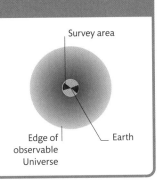

Survey area

Earth

Edge of observable Universe

The cosmic web

The Universe is not a random collection of stars and galaxies. Instead, it is a cosmic web made of connecting filaments and walls of clustered galaxies and gases stretched across the Universe, with giant voids in between, like odd-shaped bubbles. Together, these structures give the Universe a foamy appearance. However, it is thought there may be a limit on how big structures are when you zoom out far enough. This limit is sometimes called the End of Greatness.

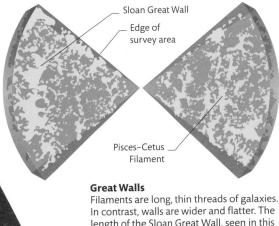

Sloan Great Wall

Edge of survey area

Pisces–Cetus Filament

Great Walls
Filaments are long, thin threads of galaxies. In contrast, walls are wider and flatter. The length of the Sloan Great Wall, seen in this survey image, is about one-sixtieth of the diameter of the observable Universe.

Sheetlike structures are known as walls

Voids contain no galaxies or only a few and have less than 10 percent of the Universe's average matter density

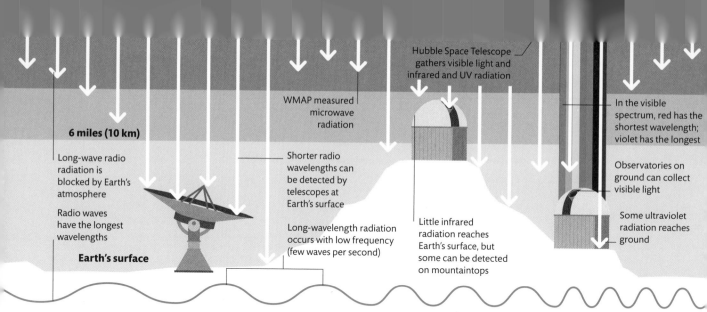

6 miles (10 km)

Long-wave radio radiation is blocked by Earth's atmosphere

Radio waves have the longest wavelengths

Earth's surface

WMAP measured microwave radiation

Shorter radio wavelengths can be detected by telescopes at Earth's surface

Long-wavelength radiation occurs with low frequency (few waves per second)

Hubble Space Telescope gathers visible light and infrared and UV radiation

Little infrared radiation reaches Earth's surface, but some can be detected on mountaintops

In the visible spectrum, red has the shortest wavelength; violet has the longest

Observatories on ground can collect visible light

Some ultraviolet radiation reaches ground

Radio waves
Stars and galaxies, as well as radio galaxies, quasars, pulsars, and masers, are all radio sources.

Microwaves
The background radiation lingering from the Big Bang is detected as microwaves.

Infrared
Infrared is heat. It can reveal dim galaxies, brown dwarfs, nebulae, and interstellar molecules.

Visible light
Emitted by most stars and some nebulae and reflected by planets and clouds, light is a rich data source.

Light

Light is the electromagnetic radiation we detect with our eyes. All forms of matter emit electromagnetic radiation, and we know about the Universe by studying radiation from distant objects such as stars.

Light in space

All kinds of radiation, including light, travel through space in straight lines at the same incredible speed—186,282 miles (299,792 km) per second—although with different wavelengths, depending on its energy. Light has no mass but can still be absorbed, reflected, or refracted when it meets something—and its path can be bent by the curved space created by a strong gravitational field (see pp.154–155). As light radiates from a source, it spreads out and its power diminishes, which is why distant galaxies appear faint.

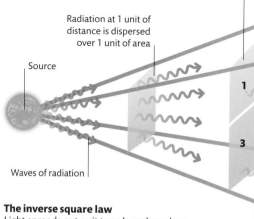

As distance from source has doubled, intensity of radiation has dropped by a factor of four

Radiation at 1 unit of distance is dispersed over 1 unit of area

Source

Waves of radiation

Radiation at 2 units of distance is dispersed over 4 units of area

The inverse square law

Light spreads out as it travels and weakens according to a rule called the inverse square law. As the distance from a source doubles, the light is spread over four times the area. Astronomers use this law to calculate distances to stars from their apparent brightness.

Radiation and Earth's atmosphere
Some kinds of radiation pass right through Earth's atmosphere to reach ground level. Others are absorbed by the atmosphere to various extents and can be detected only from space or at high altitude.

Chandra X-ray Observatory uses mirrors to focus X-rays and then produce images

Fermi telescope detects gamma-ray bursts

Tanks of ultra-pure water can detect radiation from gamma-ray bursts

Wavelength is the distance from one peak to the next

Short-wavelength radiation occurs with high frequency (many waves per second)

Gamma rays have the shortest wavelengths

Ultraviolet (UV)
UV is emitted by hot sources, including white dwarfs, neutron stars, and Seyfert galaxies, but it cannot penetrate Earth's atmosphere.

X-rays
X-rays are useful for detecting binary star systems, black holes, neutron stars, galaxy collisions, hot gases, and more.

Gamma rays
Gamma rays reveal high-energy activity from solar flares, neutron stars, black holes, exploding stars, and supernova remnants.

The electromagnetic spectrum

Light is the radiation in just one wavelength band in the huge range of wavelengths called the electromagnetic spectrum. At one end are long, low-frequency waves—radio waves, microwaves, and infrared light. At the other are short, high-frequency waves—ultraviolet light, X-rays, and gamma rays. Stars and galaxies emit all these waves in different amounts. Although the human eye can see only visible light, telescopes that can detect other wavelengths can tell us much more.

GAMMA-RAY RADIATION IS MORE THAN 100,000 TIMES MORE ENERGETIC THAN VISIBLE LIGHT

CAN ANYTHING TRAVEL FASTER THAN THE SPEED OF LIGHT?

No. According to Albert Einstein's special theory of relativity, the speed of light is the upper speed limit for ordinary matter and radiation.

PARTICLE OR WAVE?

Light and other kinds of electromagnetic radiation are emitted as packets of energy called photons. A photon is the smallest possible discrete packet, or quantum, of radiation. Photons can be understood as either particles or waves, depending on how they are encountered. This two-fold nature of light is referred to as wave-particle duality.

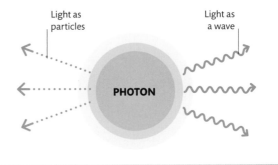

Light as particles

Light as a wave

PHOTON

Space-time

In space-time, the three dimensions of space join with time to make a 4D grid. This idea reveals how objects move through time, as well as space. It has also changed our understanding of gravity.

What is space-time?

In space-time, time and space are inseparably joined to form a grid that scientists often liken to a sheet of rubber. The sheet has two dimensions but represents four-dimensional space-time and shows bends in time, as well as space. In his general theory of relativity, Albert Einstein showed how space-time is warped around objects with mass. The more massive the object, the greater the distortion. This warping controls how everything in the Universe moves, even light. Gravity, Einstein realized, is simply the effect of these distortions on the way things move.

Objects move along imaginary lines called geodesics representing shortest distances between points in space-time

Flexible sheet representing space-time

APPROACHING COMET

In space warped by mass, geodesics curve; an object moving along a geodesic, such as a planet orbiting the Sun, will change direction due to gravity

EARTH'S ORBIT

Curvature of space means Earth is falling toward the Sun, but inertia stops it from falling into the Sun; this means Earth orbits in a curved path around the Sun

EARTH

Curved space-time
The huge mass of the Sun warps space-time around it like a heavy ball on a rubber sheet. Objects moving through its gravitational field, such as Earth, comets, and even light, are bent toward it.

Gravitational waves

In 1916, Einstein predicted that massive, accelerating objects might send out ripples in the fabric of space-time. Scientists now think that these ripples, known as gravitational waves, are set off by cataclysmic events in space—such as supernovae and colliding neutron stars and black holes—and that they travel away from their sources at the speed of light. Although they are hard to detect, gravitational waves may in future provide an alternative to electromagnetic radiation as a way of seeing things in space, such as black holes and dark matter.

Ripples from black holes
The existence of gravitational waves was confirmed in 2015, when ripples from two black holes colliding 1.3 billion light-years away were picked up on Earth using a technique called laser interferometry.

Fast-moving black holes make ripples (waves) in space-time

Black holes move gradually faster and come closer together

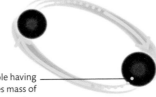

Black hole having 20 times mass of Sun but occupying much smaller space

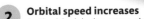

1 **Colliding black holes**
The two black holes were the remnants of collapsed giant stars. As they came close, they orbited each other for maybe millions of years before causing significant ripples.

2 **Orbital speed increases**
As the black holes came closer, they began to send gravitational waves out through the surrounding space-time. This released energy, allowing them to orbit closer and faster.

Fast-moving comet moves toward the Sun as it enters curved space-time

Light beams are also deflected by warped space-time; a beam from a star curves, and so the light appears to come from a different part of the sky

ACTUAL POSITION OF STAR

Distance between geodesics increases near massive object

APPARENT POSITION OF STAR

SUN

Up close, geodesics appear straight

The Sun is the largest object in the Solar System, so motions of all other objects in the area are affected by how it warps space

Light detected on Earth seems to have come from a point in a straight line from observer

DOES TIME ALWAYS PASS AT THE SAME RATE?

No. A rapidly moving clock ticks more slowly than a clock at rest. A clock on a spaceship traveling at 87 percent of the speed of light will tick at half the speed of a clock on Earth.

THE **APOLLO MISSIONS** WERE PLANNED USING **ISAAC NEWTON'S LAWS** OF MOTION AND GRAVITATION RATHER THAN **EINSTEIN'S**

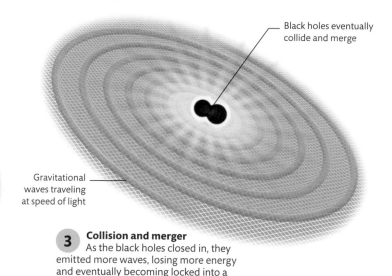

Black holes eventually collide and merge

Gravitational waves traveling at speed of light

3 **Collision and merger**
As the black holes closed in, they emitted more waves, losing more energy and eventually becoming locked into a runaway collision. The final crunch sent massive shockwaves out through space-time.

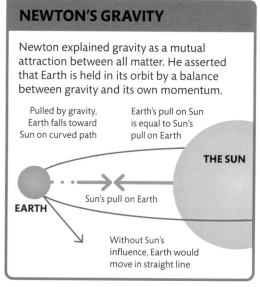

NEWTON'S GRAVITY

Newton explained gravity as a mutual attraction between all matter. He asserted that Earth is held in its orbit by a balance between gravity and its own momentum.

Pulled by gravity, Earth falls toward Sun on curved path

Earth's pull on Sun is equal to Sun's pull on Earth

THE SUN

EARTH

Sun's pull on Earth

Without Sun's influence, Earth would move in straight line

Looking back in time

When we look into space, the stars and galaxies we see are vast distances away. Looking at them means we are also looking back in time, seeing them as they were when the light left them.

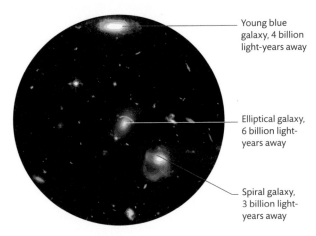

Looking into deep space
The Hubble Deep Field images of galaxies billions of light-years away reveal how the galaxies appeared billions of years ago.

- Young blue galaxy, 4 billion light-years away
- Elliptical galaxy, 6 billion light-years away
- Spiral galaxy, 3 billion light-years away

Lookback time

Although light moves faster than anything else in the Universe—at about 190,000 miles (300,000 km) per second—it does not reach us instantaneously. The farther away an object is, the longer light takes to reach us, so the farther back in time we are seeing. An object's lookback, or time-travel, distance (see pp.160–161) is also a measure of how long its light has been traveling to us—its lookback time.

How far away in time and space?
Even light from nearby objects, such as those in the Solar System, takes an appreciable time to reach us. Light takes more than eight minutes to arrive from the Sun and 1.3 seconds from the Moon.

Seeing into deep time

One of the most distant objects readily visible to the naked eye is the Andromeda Galaxy. It is about 2.5 million light-years away, which means that we see it as it was 2.5 million years ago. With the Hubble Space Telescope, we can see objects billions of light-years away and therefore as they were billions of years ago. Light from such distant objects has been red-shifted (see p.159), so it may only be possible to observe them in the infrared part of the spectrum.

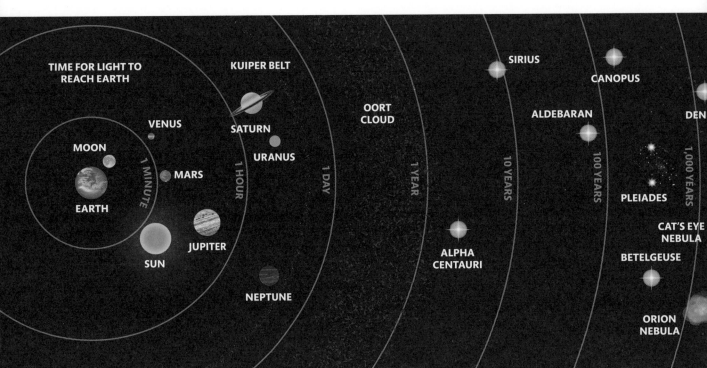

TIME FOR LIGHT TO REACH EARTH

KUIPER BELT

SIRIUS

CANOPUS

OORT CLOUD

ALDEBARAN

DEN

VENUS

SATURN

URANUS

MOON

1 MINUTE

MARS

1 HOUR

1 DAY

1 YEAR

10 YEARS

100 YEARS

1,000 YEARS

PLEIADES

EARTH

CAT'S EYE NEBULA

BETELGEUSE

JUPITER

ALPHA CENTAURI

SUN

NEPTUNE

ORION NEBULA

THE EARLIEST MOMENTS OF THE UNIVERSE

Although we cannot directly observe the earliest moments of the Universe, we can investigate what they might have been like by using particle accelerators (such as the Large Hadron Collider) to smash together subatomic particles and recreate the conditions that are thought to have existed immediately after the Big Bang.

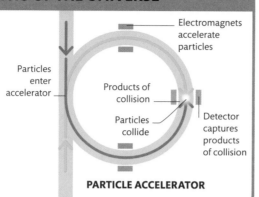

Electromagnets accelerate particles

Particles enter accelerator

Products of collision

Particles collide

Detector captures products of collision

PARTICLE ACCELERATOR

GALAXY GN-Z11 IS ONE OF THE MOST DISTANT OBJECTS EVER DETECTED—WE SEE IT AS IT WAS ABOUT 13.4 BILLION YEARS AGO

The limit of deep time observation

Light particles (photons) could not travel freely in the early Universe, so we cannot observe it directly. About 380,000 years after the Big Bang, in a period known as recombination (see pp.164–165), photons became able to move freely. These photons form the cosmic microwave background and are the oldest it is possible to detect.

The dark beginning
The early Universe was filled with plasma (a hot, dense "soup" of electrically charged particles) that stopped photons from moving freely.

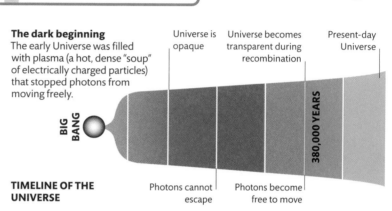

Universe is opaque

Universe becomes transparent during recombination

Present-day Universe

BIG BANG

380,000 YEARS

TIMELINE OF THE UNIVERSE

Photons cannot escape

Photons become free to move

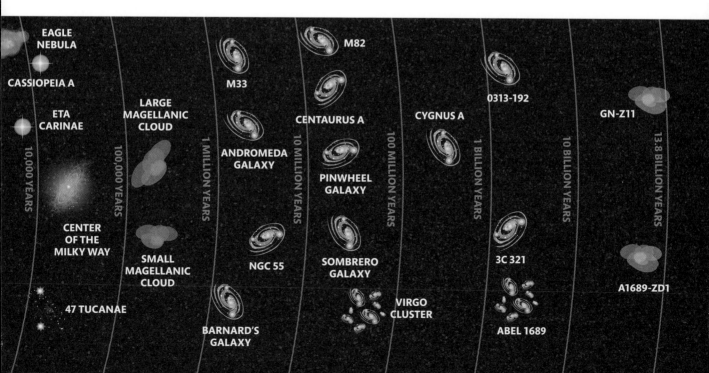

EAGLE NEBULA

CASSIOPEIA A

ETA CARINAE

LARGE MAGELLANIC CLOUD

M33

M82

0313-192

CENTAURUS A

CYGNUS A

GN-Z11

10,000 YEARS

100,000 YEARS

1 MILLION YEARS

ANDROMEDA GALAXY

10 MILLION YEARS

PINWHEEL GALAXY

100 MILLION YEARS

1 BILLION YEARS

10 BILLION YEARS

13.8 BILLION YEARS

CENTER OF THE MILKY WAY

SMALL MAGELLANIC CLOUD

NGC 55

SOMBRERO GALAXY

3C 321

A1689-ZD1

47 TUCANAE

BARNARD'S GALAXY

VIRGO CLUSTER

ABEL 1689

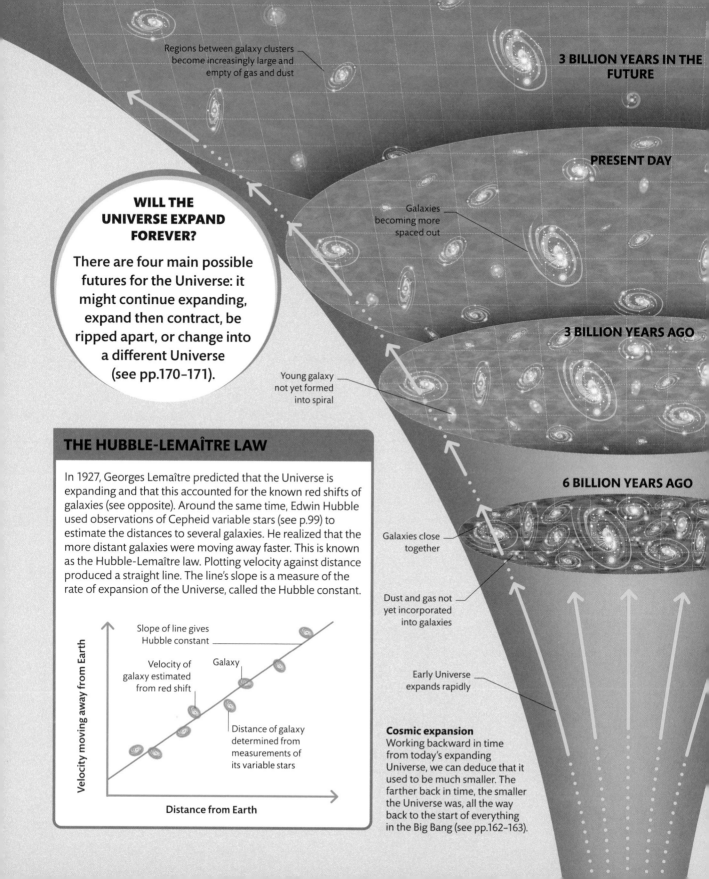

Regions between galaxy clusters become increasingly large and empty of gas and dust

3 BILLION YEARS IN THE FUTURE

PRESENT DAY

Galaxies becoming more spaced out

3 BILLION YEARS AGO

Young galaxy not yet formed into spiral

6 BILLION YEARS AGO

Galaxies close together

Dust and gas not yet incorporated into galaxies

Early Universe expands rapidly

WILL THE UNIVERSE EXPAND FOREVER?

There are four main possible futures for the Universe: it might continue expanding, expand then contract, be ripped apart, or change into a different Universe (see pp.170–171).

THE HUBBLE-LEMAÎTRE LAW

In 1927, Georges Lemaître predicted that the Universe is expanding and that this accounted for the known red shifts of galaxies (see opposite). Around the same time, Edwin Hubble used observations of Cepheid variable stars (see p.99) to estimate the distances to several galaxies. He realized that the more distant galaxies were moving away faster. This is known as the Hubble-Lemaître law. Plotting velocity against distance produced a straight line. The line's slope is a measure of the rate of expansion of the Universe, called the Hubble constant.

Slope of line gives Hubble constant

Velocity of galaxy estimated from red shift

Galaxy

Distance of galaxy determined from measurements of its variable stars

Velocity moving away from Earth

Distance from Earth

Cosmic expansion
Working backward in time from today's expanding Universe, we can deduce that it used to be much smaller. The farther back in time, the smaller the Universe was, all the way back to the start of everything in the Big Bang (see pp.162–163).

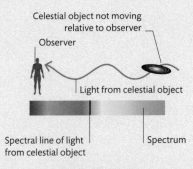

Some galaxies have evolved into spirals

Universe expanding at increasing rate

ALTHOUGH **SPACE** IS **EXPANDING,** THE **OBJECTS** WITHIN SPACE STAY THE **SAME SIZE**

The expanding Universe

Every second, the distance between objects in the Universe is getting bigger, like dots on the surface of a balloon that is being blown up. This is because the very fabric of space itself is expanding. We know that the rate of expansion is speeding up, but we do not know why or exactly how quickly.

The nature of expansion

Galaxies and other celestial objects are not moving away from each other through space. Instead, space itself is expanding and carrying the objects with it, although in localized regions, objects may move toward each other if their gravitational attraction is strong enough. There are two methods for calculating how fast the Universe is expanding: using the cosmic microwave background radiation (see pp.164–165) and measuring the red shift in the light from certain stars. The methods give different results, but a generally accepted estimate is that the Universe is expanding at about 12 miles (20 km) per second every million light-years.

Movement and wavelength
When an object and an observer are not moving relative to each other, the observer sees the true wavelength of light from the object. But if they are moving apart, the wavelength becomes longer, an effect called red shift; if they are moving closer to each other, the wavelength becomes shorter, known as blue shift.

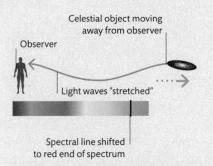

Celestial object not moving relative to observer

Observer

Light from celestial object

Spectral line of light from celestial object

Spectrum

OBSERVER AND OBJECT STATIONARY

Celestial object moving away from observer

Observer

Light waves "stretched"

Spectral line shifted to red end of spectrum

OBSERVER AND OBJECT MOVING APART

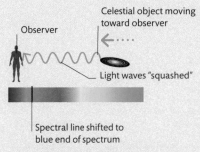

Celestial object moving toward observer

Observer

Light waves "squashed"

Spectral line shifted to blue end of spectrum

OBSERVER AND OBJECT MOVING CLOSER TOGETHER

Measuring distance

Space is expanding, so the current distance to an object in space, called its proper distance, is greater than the distance light from the object has traveled to reach us, known as the lookback distance. However, when astronomers give the distances of objects, that figure is usually the lookback one, because the exact proper distance depends on the rate of expansion of the Universe (see pp.158–159), which is uncertain.

Lookback and proper distance

The lookback distance is how far light has traveled from an object to reach us today. The proper distance is the true distance from us to the object. It is greater than the lookback distance due to the Universe's expansion.

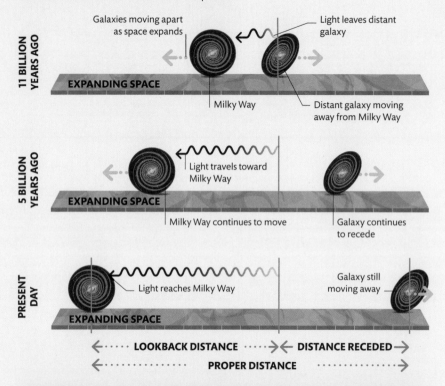

11 BILLION YEARS AGO

Galaxies moving apart as space expands

Light leaves distant galaxy

EXPANDING SPACE

Milky Way

Distant galaxy moving away from Milky Way

5 BILLION YEARS AGO

Light travels toward Milky Way

EXPANDING SPACE

Milky Way continues to move

Galaxy continues to recede

PRESENT DAY

Light reaches Milky Way

Galaxy still moving away

EXPANDING SPACE

← · · · · · LOOKBACK DISTANCE · · · · · → ← DISTANCE RECEDED →
← · · · · · · · · · · · · PROPER DISTANCE · · · · · · · · · · · · →

(see pp.158–159)

HOW BIG IS THE UNIVERSE?

The Universe is bigger than the part we can observe. We do not know exactly how much bigger, but some models estimate that it could be a sphere at least 7 trillion light-years across.

Current distance from Earth of the most distant visible objects in the Universe that are theoretically visible

Region beyond observable Universe

THE FARTHEST VISIBLE GALAXY

Detected by the Hubble Space Telescope in 2016, GN-z11 is the most distant galaxy observed from Earth. Formed about 400 million years after the Big Bang, it is located at a lookback distance of about 13.4 billion light-years. During the time taken for its light to reach us, the Universe has expanded and GN-z11 is now at a proper distance from Earth estimated to be 32 billion light-years.

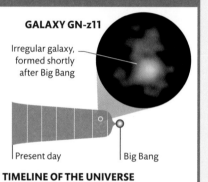

GALAXY GN-z11

Irregular galaxy, formed shortly after Big Bang

Present day

Big Bang

TIMELINE OF THE UNIVERSE

How far can we see?

The Universe is expanding and has been since its beginning in the Big Bang. This means there is a huge region, possibly infinitely large, that we cannot see because light has not had enough time to reach us from those distant parts.

The observable universe

Extending 46.5 billion light-years from Earth in every direction is a region of space called the observable Universe. This spherical region makes up every part of the Universe we have the potential to see, because light has had enough time (the age of the Universe, or 13.8 billion years) to reach us. The size of the observable Universe does not depend on the ability of our technology to detect distant objects. Instead, it is a limit resulting from the Universe's age and the finite speed of light, both of which are fundamental physical properties that cannot be overcome.

The observable sphere

Centered on Earth, the observable Universe is a spherical volume of space about 93 billion light-years in diameter. We can see objects that have a proper distance of more than 13.8 billion light-years, because the Universe has expanded while light has been traveling from them.

Outer edge of the observable Universe, called the cosmic light horizon

GN-z11—the farthest known galaxy (estimated proper distance: 32 billion light-years)

SN 1000+0216—the farthest known supernova (estimated proper distance: 23 billion light-years)

46.5 BILLION LIGHT-YEARS

ULAS J1342+0928—the farthest known quasar (estimated proper distance: 29 billion light-years)

Icarus (MACS J1149 Lensed Star 1)—the farthest known star (estimated proper distance: 14.4 billion light-years)

13.8 BILLION LIGHT-YEARS

EARTH

Distance that light has traveled from the most distant objects that are theoretically visible—the maximum lookback distance of observable objects

EDGE OF OBSERVABLE UNIVERSE

LIGHT FROM ANYTHING **MORE THAN 60 BILLION LIGHT-YEARS** AWAY WILL **NEVER REACH EARTH**

The Big Bang

Today, the Universe is teeming with stars, planets, and galaxies, but it started life about 13.8 billion years ago as an infinitely tiny speck that began expanding and is still growing.

The beginning

Wind back the expansion of the Universe and everything gets crammed into a very small space—a singularity. This super-hot, super-dense beginning is called the Big Bang. In the first fractions of a second, the singularity grew at faster than light-speed in a period known as inflation, at the end of which the Universe consisted of a sea of particles and antiparticles. The Universe then continued to expand, but at a slower rate, and eventually developed into the cosmos we are familiar with today.

The birth of the Universe

The Big Bang was not an enormous explosion in space, but an incredibly fast expansion from a single point. Everything in the modern Universe was in that point, which is why astronomers say that the Big Bang happened everywhere at once.

WHAT WAS BEFORE THE BIG BANG?

The Big Bang is generally believed to be the start of everything, including time, so it makes no sense to talk about a time before time itself existed.

Sea of particles and antiparticles emerges as inflation ends

Quark

Antiquark

Gluon

Gravity, the first fundamental force, emerges

THE BIG BANG

10^{-43} SECONDS AFTER BIG BANG

10^{-36} SECONDS AFTER BIG BANG

10^{-32} SECONDS AFTER BIG BANG

ONE-TRILLIONTH OF A SECOND AFTER BIG BANG

Universe forms from infinitely small, dense, hot point—a singularity

Inflation begins and Universe expands at incredible speed

Electron

Photon

Positron

Fundamental forces

In the first instants after the Big Bang, there was only energy; matter did not exist. In the present, four fundamental forces are at work, but these were initially unified into a single superforce. The four forces soon peeled off the superforce until they had completely separated out by one-trillionth of a second (10^{-12} seconds) after the Big Bang.

IF INFLATION WAS REPEATED TODAY, A CELL WOULD GROW LARGER THAN THE OBSERVABLE UNIVERSE

The separation of forces

Physicists believe that the four fundamental forces that govern how particles interact (the strong nuclear force, electromagnetism, and gravity) and how radioactive decay occurs (the weak nuclear force) were originally one single force but separated out soon after the Big Bang, although they do not yet know how the separation occurred.

SUPER-FORCE

GRAND UNIFIED FORCE

ELECTROWEAK FORCE

STRONG NUCLEAR FORCE

WEAK NUCLEAR FORCE

ELECTROMAGNETISM

GRAVITY

SECONDS AFTER BIG BANG 10^{-43} 10^{-36} 10^{-12}

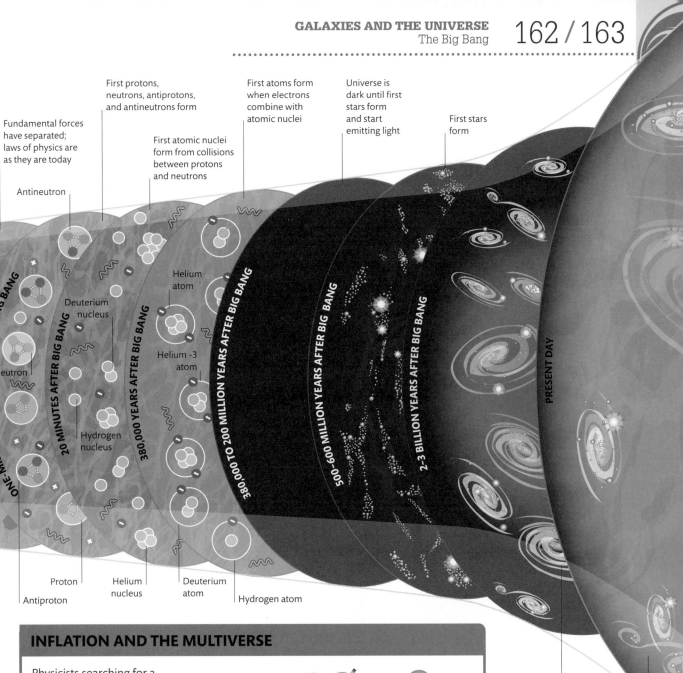

Fundamental forces have separated; laws of physics are as they are today

First protons, neutrons, antiprotons, and antineutrons form

First atomic nuclei form from collisions between protons and neutrons

First atoms form when electrons combine with atomic nuclei

Universe is dark until first stars form and start emitting light

First stars form

Antineutron

THE BIG BANG

ONE-MILL...

Neutron

Deuterium nucleus

20 MINUTES AFTER BIG BANG

Hydrogen nucleus

Helium atom

Helium -3 atom

Helium nucleus

380,000 YEARS AFTER BIG BANG

Deuterium atom

Hydrogen atom

380,000 TO 200 MILLION YEARS AFTER BIG BANG

500–600 MILLION YEARS AFTER BIG BANG

2–3 BILLION YEARS AFTER BIG BANG

PRESENT DAY

Proton

Antiproton

Some galaxies start taking on spiral shapes

Universe continues to expand

INFLATION AND THE MULTIVERSE

Physicists searching for a mechanism for inflation have found it difficult to get it to occur only once in simulations. It seems that inflation is more likely to be eternal, constantly creating new universes—a multiverse. However, so far, this idea remains controversial and there is no obvious way of testing it experimentally.

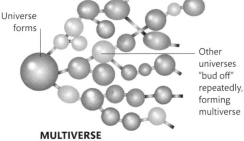

Universe forms

Other universes "bud off" repeatedly, forming multiverse

MULTIVERSE

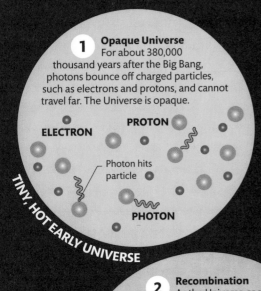

1 Opaque Universe
For about 380,000 thousand years after the Big Bang, photons bounce off charged particles, such as electrons and protons, and cannot travel far. The Universe is opaque.

ELECTRON

PROTON

Photon hits particle

PHOTON

TINY, HOT EARLY UNIVERSE

Recombination
The early Universe was too hot for protons and electrons to exist combined as atoms and too dense for photons to move freely. As the Universe expanded, it cooled and became less dense. Starting about 380,000 years after the Big Bang, in a period known as recombination, it cooled and expanded sufficiently to enable protons and electrons to combine to form hydrogen atoms and photons to travel freely.

The origin of the CMB
After recombination, the Universe was filled with small atoms (mainly hydrogen but also small amounts of helium and lithium). The atoms did not block photons (light particles) like the dense plasma did before, and they could travel freely. These photons can be detected now as the CMB radiation.

THE CMB EVERYWHERE IS AT AN AVERAGE TEMPERATURE OF -454.765°F (-270.425°C)

2 Recombination
As the Universe cools, protons and electrons combine to form atoms (mainly hydrogen). The photons are not scattered by these atoms, so the Universe becomes transparent.

HYDROGEN ATOM

PHOTON

Photon free to move

UNIVERSE COOLS AND EXPANDS

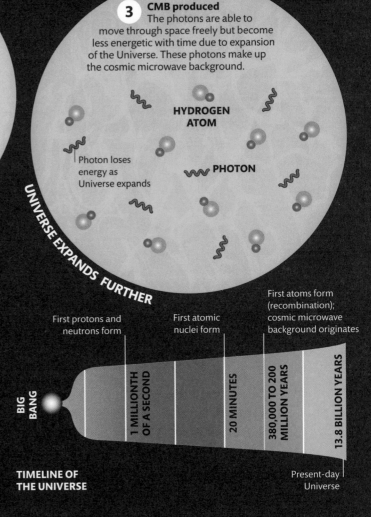

3 CMB produced
The photons are able to move through space freely but become less energetic with time due to expansion of the Universe. These photons make up the cosmic microwave background.

HYDROGEN ATOM

Photon loses energy as Universe expands

PHOTON

UNIVERSE EXPANDS FURTHER

Early radiation

The very early Universe was opaque. Light could only move freely once the first atoms had formed. The relic radiation from this period forms the cosmic microwave background (CMB) and is the earliest radiation we can detect.

First protons and neutrons form

First atomic nuclei form

First atoms form (recombination); cosmic microwave background originates

BIG BANG

1 MILLIONTH OF A SECOND

20 MINUTES

380,000 TO 200 MILLION YEARS

13.8 BILLION YEARS

TIMELINE OF THE UNIVERSE

Present-day Universe

Measuring the CMB

Since the discovery of the CMB in 1964, hundreds of experiments have been conducted to measure and study the radiation. The most complete picture was put together using data gathered by Europe's Planck space observatory, from 2009 to 2013. The CMB looks almost identical in every direction but has tiny fluctuations that differ in temperature by only a fraction of a degree. These represent differences in densities that were present right after the Universe formed. They started as tiny variations but, as the Universe expanded, the fluctuations grew along with it, and areas with higher density in the early Universe turned into huge structures like galaxy clusters.

The earliest radiation
This image, obtained by the Planck observatory, shows the whole sky projected onto a flat surface. The temperature variations relate to irregularities in the density of matter in the early Universe. Areas of higher-than-average temperature indicate areas of higher density, and vice versa.

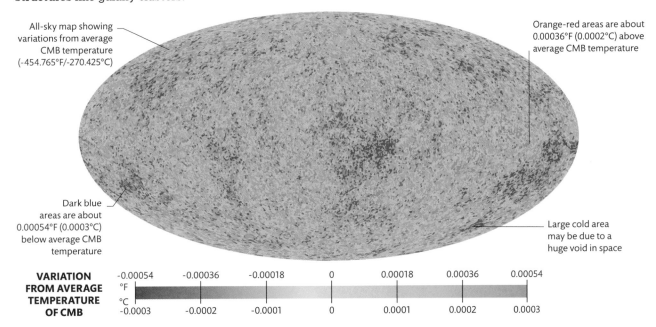

All-sky map showing variations from average CMB temperature (-454.765°F/-270.425°C)

Orange-red areas are about 0.00036°F (0.0002°C) above average CMB temperature

Dark blue areas are about 0.00054°F (0.0003°C) below average CMB temperature

Large cold area may be due to a huge void in space

VARIATION FROM AVERAGE TEMPERATURE OF CMB							
-0.00054 °F	-0.00036	-0.00018	0	0.00018	0.00036	0.00054	
-0.0003 °C	-0.0002	-0.0001	0	0.0001	0.0002	0.0003	

OTHER EVIDENCE FOR THE BIG BANG THEORY

The existence of the cosmic microwave background radiation provides strong evidence in support of the Big Bang theory of the origin of the Universe. Other observations also provide support for the theory.

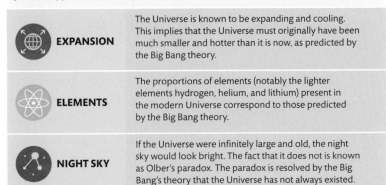

	EXPANSION	The Universe is known to be expanding and cooling. This implies that the Universe must originally have been much smaller and hotter than it is now, as predicted by the Big Bang theory.
	ELEMENTS	The proportions of elements (notably the lighter elements hydrogen, helium, and lithium) present in the modern Universe correspond to those predicted by the Big Bang theory.
	NIGHT SKY	If the Universe were infinitely large and old, the night sky would look bright. The fact that it does not is known as Olber's paradox. The paradox is resolved by the Big Bang's theory that the Universe has not always existed.

WHY IS THE CMB SO COLD?

Originally, the CMB had a much shorter wavelength and higher energy, corresponding to about 5,000°F (3,000°C). As the Universe expanded, the radiation was stretched to longer wavelengths, which have less energy and so are colder.

Early particles

Shortly after the Big Bang, the first particles emerged from a sea of energy. They would go on to form the building blocks of the modern Universe.

The first nuclei

Initially, the Universe was inconceivably hot, and matter and energy were in an interchangeable form known as mass-energy. As the cosmos cooled, fundamental particles, including quarks (see opposite), emerged. The strong nuclear force (see p.162) bound quarks together to form protons and neutrons, which make up the nuclei of all atoms.

The origin of matter

By the time the Universe was just 20 minutes old, the first atomic nuclei had formed. Matter and antimatter (see opposite) were both present, in the form of particles and antiparticles.

TIMELINE OF THE UNIVERSE

Particles and antiparticles form, then annihilate each other, creating energy and leaving small residue of matter particles

First protons and neutrons form

First atoms form (recombination)

BIG BANG

10^{-32}–10^{-9} SECONDS

1 MILLIONTH OF A SECOND

20 MINUTES

380,000 TO 200 MILLION YEARS

13.8 BILLION YEARS

First atomic nuclei form

Present-day Universe

THE **HYDROGEN NUCLEI** IN A GLASS OF WATER WERE CREATED IN **THE FIRST FEW MINUTES** OF THE UNIVERSE'S LIFE

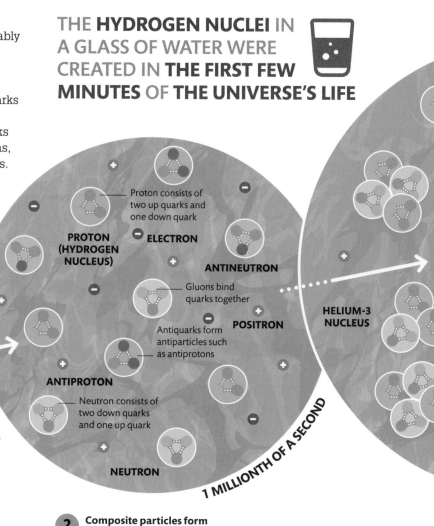

PROTON (HYDROGEN NUCLEUS)

Proton consists of two up quarks and one down quark

ELECTRON

ANTINEUTRON

Gluons bind quarks together

POSITRON

Antiquarks form antiparticles such as antiprotons

ANTIPROTON

Neutron consists of two down quarks and one up quark

NEUTRON

HELIUM-3 NUCLEUS

GLUON
ELECTRON
UP QUARK
DOWN QUARK
POSITRON
DOWN ANTIQUARK
UP ANTIQUARK

10^{-32}–10^{-9} SECONDS

1 MILLIONTH OF A SECOND

1 Particles and antiparticles form
The first quarks and antiquarks formed spontaneously from the sea of mass-energy during a fleeting period called the quark epoch. The first electrons and positrons also emerged in a process known as leptogenesis.

2 Composite particles form
Quarks were bound together by gluons, which carry the strong nuclear force, to form protons and neutrons, which are both types of composite particle. A proton has an overall positive electrical charge; neutrons have no charge.

The first atoms

An atom comprises a positively charged nucleus surrounded by one or more negatively charged electrons, held together by the electromagnetic force. The first nuclei formed within minutes of the Big Bang, but it was 380,000 years before the Universe had cooled enough for them to join with electrons in the process of recombination (see p.164) to make atoms of the first three elements.

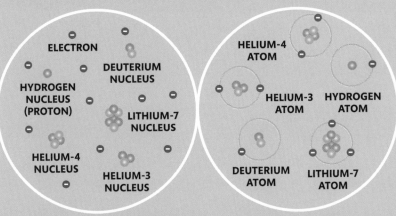

1 **Separate nuclei and electrons**
For many hundreds of thousands of years, atomic nuclei and electrons existed separately in a hot plasma of fast-moving particles.

2 **Atoms form**
Eventually, electrons were captured by atomic nuclei to form atoms of helium, hydrogen, deuterium (a heavy form of hydrogen), and lithium.

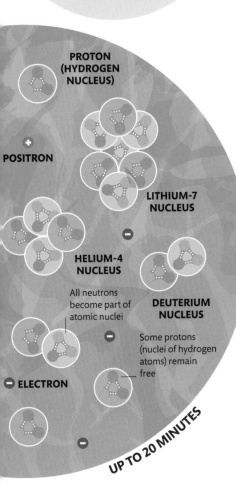

3 **Nuclei form**
Collisions between protons and neutrons formed the nuclei of some atoms, including helium-4, and small amounts of other nuclei, such as helium-3, deuterium, and lithium-7.

SUBATOMIC PARTICLES

Atoms are made up of smaller, subatomic particles—protons, neutrons, and electrons. Electrons are fundamental particles, which means they are not made of smaller particles. But protons and neutrons are both made of fundamental particles known as quarks and gluons. Each particle has a corresponding antiparticle.

UP QUARK   DOWN QUARK

ELECTRON   GLUON

PHOTON   HIGGS BOSON

Fundamental particles
Some of these, such as quarks, are building blocks of matter. Others, such as gluons and photons, are force-carriers.

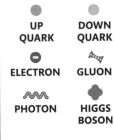

PROTON   NEUTRON

Composite particles
These are made up of smaller, fundamental particles, such as quarks and gluons.

UP ANTIQUARK   DOWN ANTIQUARK

POSITRON

ANTIPROTON   ANTINEUTRON

Antiparticles
Antiparticles have the same mass as their equivalent particles but exactly opposite values of other properties, including electrical charge.

The first stars and galaxies

The first stars started to form only about 200 million years after the Big Bang. The earliest galaxies began to form shortly afterward as dark matter helped clump stars together into groups. When these infant galaxies merged, it triggered yet more star formation.

The first stars

Early in the life of the Universe, the only ingredients available for star formation were the hydrogen and helium made shortly after the Big Bang—the first stars contained no heavy elements. These fledgling stars were massive, dozens of times more massive than our own Sun. The intense ultraviolet light they emitted ripped electrons from hydrogen atoms, ionizing the gas between the first dwarf galaxies. The first stars died young, exploding as cataclysmic supernovae within a few million years and creating the first heavy elements.

DID THE FIRST STARS HAVE PLANETS?

The first stars may have had planets, but they would not have been rocky, because the early Universe consisted only of gas and hot plasma (a "soup" of electrically charged particles).

TIMELINE OF THE UNIVERSE

BIG BANG

First stars form 200 million years after Big Bang

First galaxies start to form 400 million years after Big Bang

Present-day Universe

380,000–400 MILLION YEARS

13.8 BILLION YEARS

Reionization begins 350 million years after Big Bang

Early star and galaxy formation
The first stars formed early in the Universe but were short-lived. The first galaxies were small and evolved into those we can see today.

Big Bang

Early Universe filled with electrically charged hydrogen and helium nuclei

First atoms start to form 380,000 years after Big Bang

Universe is filled with neutral hydrogen and helium atoms

Hydrogen and helium gases begin to clump together to form clouds

Early stars form inside gas clouds about 200 million years after Big Bang

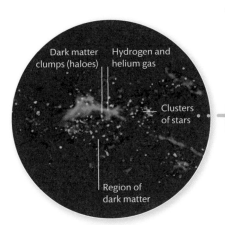

Dark matter clumps (haloes)

Hydrogen and helium gas

Clusters of stars

Region of dark matter

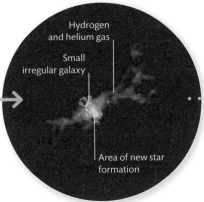

Hydrogen and helium gas

Small irregular galaxy

Area of new star formation

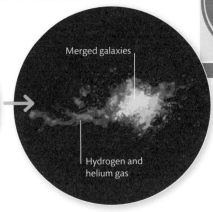

Merged galaxies

Hydrogen and helium gas

1 Dark matter clumps together
Gravitational attraction pulls dark matter together into clumps called haloes. These haloes attract normal matter, such as hydrogen and helium gas, which become compressed further.

2 Small galaxies form
Matter continues to clump together, eventually forming small irregular galaxies. Inside these galaxies, knots of denser matter develop, creating regions where new stars can start to form.

3 Galaxies merge
The galaxies, which are mostly empty space, thread through each other, creating larger galaxies and even more areas for star formation. Every major galaxy in today's Universe has undergone at least one merger.

The birth of galaxies

The processes by which the first galaxies formed are still uncertain. However, it is thought that, in the early life of the Universe, some regions of space were slightly denser than others. These denser regions attracted dark matter, which in turn pulled in gas and stars. This process continued until the first primitive galaxies formed. The galaxies we see today, such as spirals, would only form later through mergers of the more primitive galaxies.

OUR **MILKY WAY** IS ABOUT **100,000 TIMES MORE MASSIVE** THAN THE **FIRST GALAXIES**

Dwarf galaxies combine to form larger galaxies

Clusters of stars drawn together to form dwarf galaxies 400 million years after Big Bang

Ultraviolet radiation from hot stars forms bubbles of hot, electrically charged gas

Reionization starts 350 million years after Big Bang

First stars explode as supernovae 300 million years after Big Bang

Stars form in clusters that coincide with concentrations of dark matter

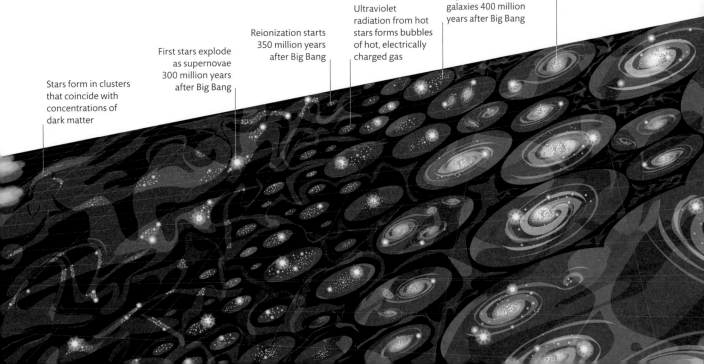

The future of the Universe

What lies in store for our cosmos depends on a battle that has been taking place since the Big Bang between gravity and a little-understood form of energy. Astronomers are still unsure of the outcome.

Dark energy

Astronomers suspect that empty space is full of a mysterious substance or force called dark energy that acts in opposition to gravity. There is always the same amount of dark energy in any given volume of space, so its potency grows as the Universe expands and space swells to a larger volume. This might explain why the expansion of the Universe is accelerating.

Possible futures

What will ultimately happen to space depends on whether the gravitational attraction between stars, galaxies, and clusters of galaxies can be overcome by dark energy. If it cannot, the Universe will collapse in on itself in a reversal of the Big Bang. Should gravity be overwhelmed, the Universe will continue expanding, potentially at a catastrophic rate. Alternatively, a new theory in physics could change all our ideas about the potential outcome.

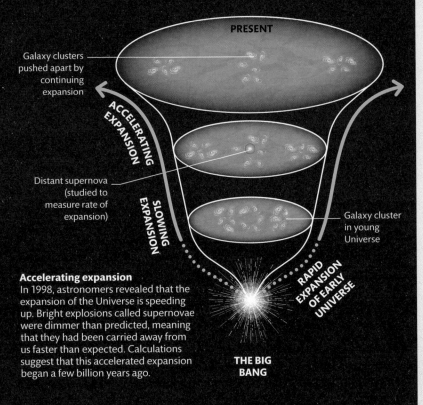

PRESENT

Galaxy clusters pushed apart by continuing expansion

ACCELERATING EXPANSION

Distant supernova (studied to measure rate of expansion)

SLOWING EXPANSION

Galaxy cluster in young Universe

RAPID EXPANSION OF EARLY UNIVERSE

THE BIG BANG

Accelerating expansion

In 1998, astronomers revealed that the expansion of the Universe is speeding up. Bright explosions called supernovae were dimmer than predicted, meaning that they had been carried away from us faster than expected. Calculations suggest that this accelerated expansion began a few billion years ago.

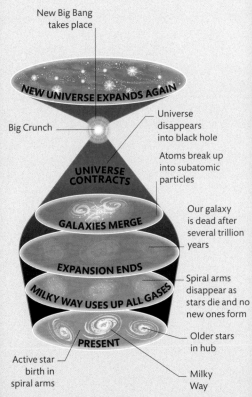

New Big Bang takes place

NEW UNIVERSE EXPANDS AGAIN

Big Crunch

Universe disappears into black hole

Atoms break up into subatomic particles

UNIVERSE CONTRACTS

Our galaxy is dead after several trillion years

GALAXIES MERGE

EXPANSION ENDS

Spiral arms disappear as stars die and no new ones form

MILKY WAY USES UP ALL GASES

Older stars in hub

PRESENT

Active star birth in spiral arms

Milky Way

The Big Crunch

This scenario would see gravity win out. The Universe would become smaller and hotter, eventually shrinking back down to a tiny speck—possibly followed by a new Big Bang. This was once a popular idea, but it has fallen from favor with the discovery of dark energy.

IN THE **DISTANT FUTURE,** THE **UNIVERSE** COULD BE **COLD AND DEAD** OR EVEN **RIPPED APART**

HOW MUCH LONGER WILL THE UNIVERSE LAST?

According to most likely scenarios, the Universe will last for billions of years and might even last for ever. However, it is theoretically possible that it could end at any time if the Big Change model is correct.

THE COSMOLOGICAL CONSTANT

The cosmological constant was introduced by Albert Einstein as an "anti-gravity" force to counterbalance the attractive force of gravity. The discovery that the expansion of the Universe is speeding up seems to imply that the cosmological constant is similar to dark energy, which tends to accelerate expansion.

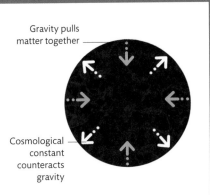

Gravity pulls matter together

Cosmological constant counteracts gravity

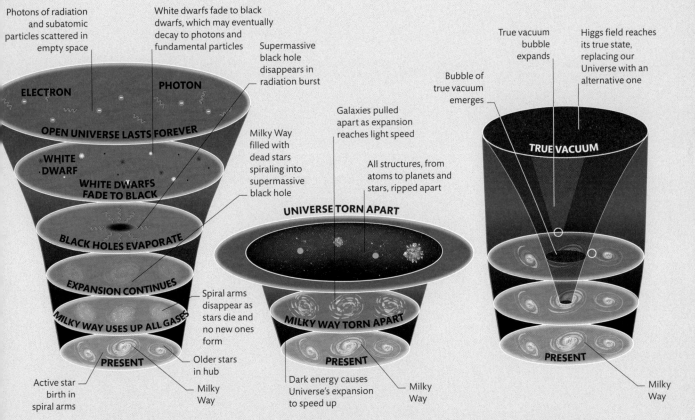

Photons of radiation and subatomic particles scattered in empty space

White dwarfs fade to black dwarfs, which may eventually decay to photons and fundamental particles

Supermassive black hole disappears in radiation burst

True vacuum bubble expands

Higgs field reaches its true state, replacing our Universe with an alternative one

Bubble of true vacuum emerges

Galaxies pulled apart as expansion reaches light speed

ELECTRON   PHOTON

OPEN UNIVERSE LASTS FOREVER

Milky Way filled with dead stars spiraling into supermassive black hole

All structures, from atoms to planets and stars, ripped apart

TRUE VACUUM

WHITE DWARF

WHITE DWARFS FADE TO BLACK

UNIVERSE TORN APART

BLACK HOLES EVAPORATE

EXPANSION CONTINUES

Spiral arms disappear as stars die and no new ones form

MILKY WAY USES UP ALL GASES

MILKY WAY TORN APART

PRESENT   Older stars in hub   Milky Way

Active star birth in spiral arms

PRESENT

Dark energy causes Universe's expansion to speed up

Milky Way

PRESENT

Milky Way

The Big Chill
If the Universe continues to expand steadily, then eventually energy and matter will become so diluted that there will not be any planets, stars, or galaxies left. Temperatures will fall to absolute zero and a sea of atomic shrapnel will be all that is left.

The Big Rip
If dark energy continues to accelerate, the expansion of the Universe, after 22 billion years or so all structures, including black holes, will be ripped apart. Even the spaces between atoms and subatomic particles will have stretched so far that they are torn apart.

The Big Change
This theory involves the Higgs boson particle and an energy field called the Higgs field. If the Higgs field reaches its lowest energy, or vacuum state, a bubble of vacuum energy could appear and expand at close to light speed, destroying everything in its path.

SPACE

EXPLORATION

Getting into space

Beyond the protective layers of Earth's atmosphere lies the vastness of outer space. The first hurdle to overcome when exploring space is simply reaching it. Overpowering the pull of Earth's gravity and achieving sufficient speed to enter a stable path around Earth, called an orbit, is the initial challenge. In order to explore interplanetary space beyond Earth's orbit, a further boost of speed and thrust is required.

A **GERMAN V-2 ROCKET** BECAME THE **FIRST OBJECT MADE BY HUMANS TO REACH SPACE,** IN 1942

Where is space?

As Earth's atmosphere gets thinner at higher altitudes, aircraft find it harder to generate lift using the pressure of air flowing under their wings. Without the molecules contained within an atmosphere to reflect or scatter light, space appears black to our perception. Outer space is generally agreed to be the region where a vehicle must enter orbit around Earth in order to remain above the surface, but there is no officially agreed upon definition for the "edge of space." US space agency NASA puts the beginning of space at 50 miles (80 km) above sea level, while the International Aeronautical Federation (FAI) puts it at 60 miles (100 km).

Exosphere

In the outermost layer of the atmosphere, beginning about 370 miles (600 km) above the surface, air pressure no longer falls with increasing altitude. The exosphere's sparse gases merge gradually into space.

Satellites orbit Earth in exosphere, where they experience only a small amount of drag

EXOSPHERE (370+ MILES/600+ KM)

THERMOSPHERE (370 MILES/600 KM)

Aurorae occur at varying altitudes, mostly in thermosphere

Low-orbiting spacecraft and space stations orbit in thermosphere

MESOSPHERE

Thermosphere

Above about 53 miles (85 km), ultraviolet radiation breaks gas molecules apart into electrically charged ions, creating a layer of hot but tenuous gas called the thermosphere. Aurorae are mostly formed in this layer.

HAS ANYONE EVER REACHED SPACE IN AN AIRPLANE?

Yes. In the 1960s, eight US pilots reached the edge of space in a hypersonic, rocket-boosted plane called the X-15, dropped by a large carrier aircraft.

Mesosphere
Above about 30–40 miles (50–65 km), atmospheric temperatures fall again within a layer called the mesosphere. This layer is too high for conventional aircraft to reach but too low for spaceflight.

(53 MILES/85 KM)

Most shooting stars burn up in mesosphere

Commercial airliners cruise troposphere

STRATOSPHERE (30 MILES/50 KM)

Highest weather balloons reach lower mesosphere

TROPOSPHERE (4–12 MILES/6–20 KM)

Stratosphere
While temperatures fall with increasing altitude in the troposphere, they increase with altitude through the stratosphere, where gases including ozone absorb the Sun's ultraviolet rays.

Troposphere
The lowest layer of Earth's atmosphere contains 75 percent of its mass and 99 percent of all its water vapor. It extends to around 12 miles (20 km) above the equator but just 4 miles (6 km) above the poles.

ESCAPING EARTH'S GRAVITY

In order to completely escape Earth's pull, a vehicle must reach a speed known as escape velocity, where it is traveling so fast that Earth's gravity can never fully slow it down. Escape velocity at Earth's surface is approximately 7 miles (11.2 km) per second, which is far greater than the speed required to achieve orbit.

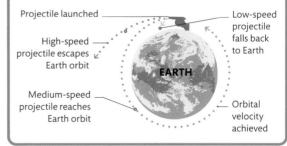

Projectile launched

Low-speed projectile falls back to Earth

High-speed projectile escapes Earth orbit

EARTH

Medium-speed projectile reaches Earth orbit

Orbital velocity achieved

Reaching orbit

In order to remain in space and not fall back to Earth, any vehicle must achieve a stable orbit—a circular or elliptical loop around Earth at sufficient height for it to avoid being slowed too much by drag from the upper atmosphere. An orbit is a path on which an object's momentum (which gives it a tendency to continue moving in a straight line) is exactly countered by the pull of gravity toward Earth. For a circular low Earth orbit (LEO) 125 miles (200 km) above the surface, this requires a spacecraft or space station to reach a speed of 17,400 miles (28,000 km) per hour.

Falling indefinitely
A really powerful throw or launch will mean that Earth's surface starts to curve away from a falling object before it can make contact with land. The object will fall "toward" Earth indefinitely, causing it to circle, or orbit, the planet repeatedly. This kind of motion is called free fall.

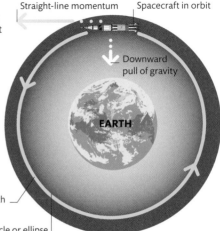

Straight-line momentum

Spacecraft in orbit

Downward pull of gravity

EARTH

Resulting curved path

Orbit may be a circle or ellipse

Rockets

Rockets are the only practical means of putting large objects into space using modern technology. **Although a rocket is simply any projectile that flies by the principle of action and reaction, a space launch requires a rocket that generates sufficient thrust to overcome the pull of gravity.**

How rockets work

Rockets are based on action and reaction. For any self-contained object, a force generated in one direction must be balanced by an equal force in the opposite direction. To generate large amounts of thrust, rockets burn chemicals called propellants. The exhaust gases escape at high speed through specially shaped nozzles, creating a reaction force that pushes the rocket in the opposite direction.

ROCKET PROPELLANTS

Rockets burn propellants to generate explosive thrust. Most combine two liquid chemicals, a fuel and an oxidizer, to produce a chemical reaction. Solid-fueled rockets are much easier to manufacture. They mix both chemicals in a solid "matrix" that burns continually once ignited within a cylinder.

LIQUID FUEL

- Hot gases
- Combustion chamber
- Liquid oxidizer
- Liquid fuel

SOLID FUEL

- Combustion chamber
- Ignition point
- Open core
- Fuel matrix
- Hot gases

Inside a liquid-fueled rocket

At launch, the bulk of a rocket (such as this European Space Agency rocket known as Ariane 5) is occupied by engines and fuel tanks. The payload to be delivered into orbit is secured on top of the uppermost stage.

Aerodynamic nose fairing reduces air resistance

Fairing protects payload during launch

PAYLOAD

Ariane 5 can launch multiple payloads into orbit

Autonomous Transfer Vehicle (ATV) for deliveries to ISS

Integrated engine for ATV orbital maneuvers

Cryogenic upper stage carries liquid fuel at low temperature

Upper-stage rocket nozzle

Solid boosters each carry 262 tons (238 tonnes) of propellant

145 tons (132 tonnes) of liquid oxygen

28 tons (26 tonnes) of liquid hydrogen

LIQUID OXYGEN TANK

SOLID BOOSTER

Igniter initiates combustion

ROCKET MOMENTUM

Rocket moves in opposite direction to exhaust

Exhaust particles ejected at high velocity

FORCE OF THRUST ←

← **PULL OF GRAVITY** →

Thrust
Rockets surmount the pull of gravity by expelling gases from engines at high speed to generate upward thrust in the opposite direction.

WHO INVENTED THE SPACE ROCKET?

The first person to seriously propose using rockets for space travel was Russian teacher, physicist, inventor, and aviation engineer Konstantin Tsiolkovsky (1857–1935).

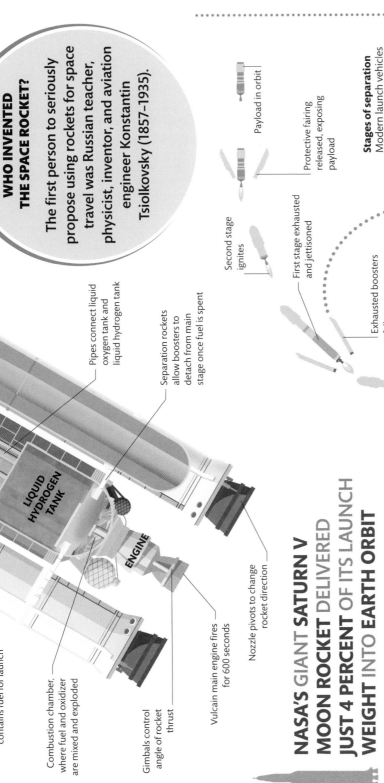

Payload in orbit

Protective fairing released, exposing payload

Second stage ignites

First stage exhausted and jettisoned

Stages of separation
Modern launch vehicles may use a cluster of small rockets around the base of a first stage, with one or more further stages above. The payload released into orbit may also be fitted with a rocket motor for further thrust and maneuverability.

Exhausted stages fall back to Earth

Exhausted boosters fall away

Lift-off achieved

First stage and boosters fire at launch

Cryogenic main stage contains fuel for launch

Pipes connect liquid oxygen tank and liquid hydrogen tank

Separation rockets allow boosters to detach from main stage once fuel is spent

Combustion chamber, where fuel and oxidizer are mixed and exploded

LIQUID HYDROGEN TANK

ENGINE

Gimbals control angle of rocket thrust

Nozzle pivots to change rocket direction

Vulcain main engine fires for 600 seconds

NASA'S GIANT SATURN V MOON ROCKET DELIVERED JUST 4 PERCENT OF ITS LAUNCH WEIGHT INTO EARTH ORBIT

Multistage rockets

Although the action and reaction forces generated in a rocket are equal, they produce a far greater acceleration in the escaping lightweight exhaust gas than they do on the mass of the rocket itself. Because the rocket must move from the outset with enough thrust to overcome gravity (to avoid falling back to Earth), it must therefore burn huge amounts of fuel in the first few moments after launch. In order to reduce the amount of excess mass carried into orbit, many rockets consist of several separate stages with separate fuel tanks and engines that are fired either in sequence or in parallel and then jettisoned as the rocket gains speed and their fuel is exhausted.

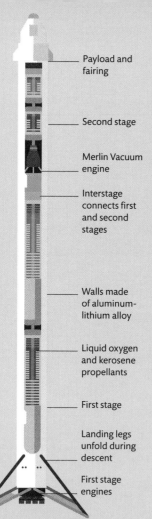

- Payload and fairing
- Second stage
- Merlin Vacuum engine
- Interstage connects first and second stages
- Walls made of aluminum-lithium alloy
- Liquid oxygen and kerosene propellants
- First stage
- Landing legs unfold during descent
- First stage engines

Reusable rockets

Traditional rockets are expensive and wasteful—not only do they burn huge amounts of fuel, but the fuel tanks and engines are also discarded and unsalvageable, despite being used on just a single flight. Developing fully reusable rockets is essential to lower the cost of access to space.

Return and recycle

Since 2015, US company SpaceX has pioneered the successful landing and reuse of rocket stages from its Falcon launch vehicles. The lower stages (either single rockets or clusters of three) are equipped with steering thrusters that guide them back to a preplanned landing site (either on land or on a floating platform at sea). They are jettisoned from the upper stage with excess fuel still on board to slow their descent during final approach.

WHAT WAS THE FIRST PARTIALLY REUSABLE SPACE VEHICLE?

The Space Shuttle, launched for the first time in 1981, featured a reusable orbiter and solid rocket boosters that could be refurbished.

Landing a rocket

With an 85 percent success rate, Falcon 9 has made the incredibly difficult task of bringing a rocket stage back to a vertical landing look deceptively simple. However, landing a rocket under power, on target, and in good condition for reuse involves some ingenious new technology.

1 Lift-off!
Falcon 9 launches vertically like any traditional rocket. The "Full Thrust" version of the rocket stands 230 ft (70 m) tall on the launchpad and consists of two stages, an interstage, and the payload with its fairing on top.

2 First-stage burn
At launch, nine Merlin engines on the rocket's first stage ignite. Arranged in a configuration known as an "octaweb," they burn a mix of RP-1 (a kerosene-based rocket fuel) and liquid oxygen.

Main engine cut-off precedes stage separation

3 Engine cut-off
The first-stage rocket engines cut out after around 180 seconds, having carried the vehicle to altitudes of around 44 miles (70 km) and speeds of around 4,400 mph (7,000 kph).

THE **MERLIN ENGINES** POWERING **FALCON 9'S FIRST STAGE** GENERATE **1.7 MILLION LB (770,000 KG) OF THRUST**

Vertical launch from launch platform

Single-stage-to-orbit vehicles

The ideal means of reaching orbit is with a single-stage-to-orbit (SSTO) vehicle that can reach space in one piece and return to Earth for a rapid turnaround. SSTO concepts include traditional vertically launched rockets, but also spaceplanes fitted with efficient hybrid engines to deliver a payload to low Earth orbit.

Inside an SSTO
The Skylon spaceplane design includes an experimental hybrid engine called SABRE to reach orbit.

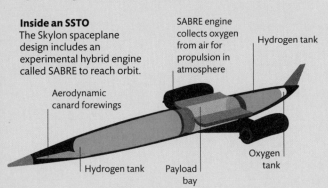

SABRE engine collects oxygen from air for propulsion in atmosphere

Hydrogen tank

Aerodynamic canard forewings

Hydrogen tank

Payload bay

Oxygen tank

SUBORBITAL FLIGHT

Blue Origin's New Shepard rocket is a vertical take-off SSTO intended to launch a passenger capsule for short flights that reach space but do not enter orbit. In November 2015, an uncrewed New Shepard was the first vertical rocket to reach space and make a return to Earth.

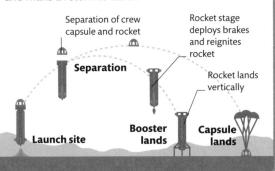

Separation of crew capsule and rocket

Rocket stage deploys brakes and reignites rocket

Separation

Rocket lands vertically

Launch site

Booster lands

Capsule lands

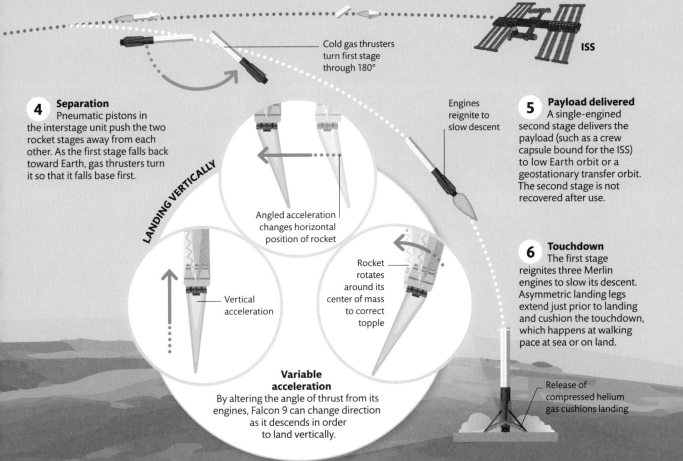

Cold gas thrusters turn first stage through 180°

ISS

4 Separation
Pneumatic pistons in the interstage unit push the two rocket stages away from each other. As the first stage falls back toward Earth, gas thrusters turn it so that it falls base first.

LANDING VERTICALLY

Angled acceleration changes horizontal position of rocket

Vertical acceleration

Rocket rotates around its center of mass to correct topple

Variable acceleration
By altering the angle of thrust from its engines, Falcon 9 can change direction as it descends in order to land vertically.

Engines reignite to slow descent

5 Payload delivered
A single-engined second stage delivers the payload (such as a crew capsule bound for the ISS) to low Earth orbit or a geostationary transfer orbit. The second stage is not recovered after use.

6 Touchdown
The first stage reignites three Merlin engines to slow its descent. Asymmetric landing legs extend just prior to landing and cushion the touchdown, which happens at walking pace at sea or on land.

Release of compressed helium gas cushions landing

Satellite orbits

A satellite's orbit is a stable circular or elliptical path around an object taken under the influence of gravity. Satellites follow a variety of orbits around Earth depending on their purpose.

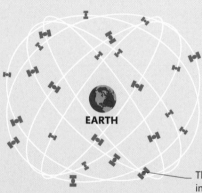

EARTH

MOLNIYA SATELLITE

The GPS constellation initially incorporated 24 orbiting satellites

Types of orbit

A satellite's speed relative to Earth's surface varies with its altitude. Those in circular orbits maintain a constant speed, with those in low orbits moving faster than those in high orbits. Elliptical orbits cause a satellite to move relatively fast at perigee (when it is closest to Earth) and slower at apogee (when it is farthest away). While some satellites orbit directly above the equator, most are inclined (tilted at an angle), so they pass over different points on the surface as Earth rotates beneath them.

Satellite constellations
Applications such as satellite telephony and navigation require multiple satellites to work together in a group known as a constellation. The satellites fly in precisely arranged low- or midaltitude Earth orbits to provide continuous coverage of Earth's surface.

GEOSTATIONARY ORBIT

Satellite follows direction of Earth's rotation

Classifying orbits
Low Earth orbits, near-circular paths in the thermosphere, are the most easily reached. Earth-observing satellites in polar orbits fly over a different band of Earth's surface on each orbit. Sun-synchronous orbits allow satellites to compare strips of Earth's surface under even lighting. Elliptical and high orbits take them much farther away from Earth, bringing more surface into view.

Geostationary orbit takes 23 hours 56 minutes to complete

COMMUNICATIONS SATELLITE

SPACE JUNK

Since the beginning of the space age in 1957, space around Earth has become increasingly crowded not only with working satellites, but also with redundant spacecraft, used rocket stages, and other debris. Collisions are a constant danger to working satellites, crewed spacecraft, and even the International Space Station and the personnel aboard.

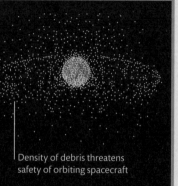

Density of debris threatens safety of orbiting spacecraft

WHAT WAS THE FIRST SATELLITE ORBIT?

Sputnik's orbit ranged from 133 to 583 miles (215 to 939 km) above the Earth and was tilted at 65° to the equator.

Orbital maneuvers

Most satellites are initially launched into low Earth orbit (LEO). From here, they use onboard engines and rocket thrusters or a final upper-stage rocket motor to reach their final desired orbit.

Changing the shape and size of orbit is far easier than altering its inclination once in space.

Transfer orbits

Satellites can move between circular orbits along paths called transfer orbits. A transfer orbit is a segment of an elliptical orbit that touches the lower circle at perigee and the upper circle at apogee. A precise engine burn is required at each stage.

Second rocket burn to enter higher circular orbit

HIGH ORBIT

LOW ORBIT

TRANSFER ORBIT

Transfer orbit carries satellite to higher altitude

North pole

Rocket thrust pushes satellite into transfer orbit

Satellite uses

The majority of satellites are designed to do specific tasks that relate to Earth. Following the right type of orbit is a vital element of getting the job done.

Satellite telephony
Satellite phone services are provided by constellations in LEOs. Several satellites are within range from any point on Earth at one moment.

Earth mapping
Sun-synchronous orbits ensure that space-based photographs of Earth's surface are all illuminated from the same direction.

Earth monitoring
Satellites designed to track various aspects of Earth's climate follow polar orbits. They can build up a complete picture of conditions on Earth.

Broadcasting
Many broadcasting satellites follow geostationary orbits over the equator, where they orbit in the same period as Earth rotates.

High latitudes
For high-latitude areas where equatorial comsats may be out of sight, satellites follow highly inclined, highly elliptical orbits called Molniya orbits.

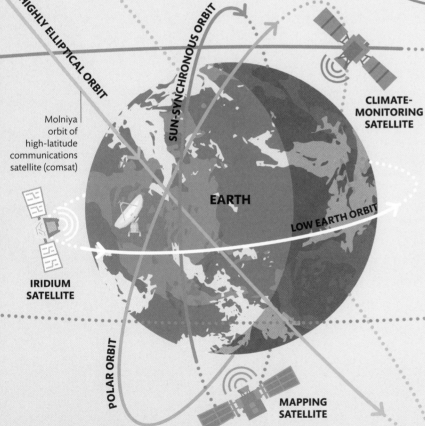

HIGHLY ELLIPTICAL ORBIT

SUN-SYNCHRONOUS ORBIT

CLIMATE-MONITORING SATELLITE

Molniya orbit of high-latitude communications satellite (comsat)

EARTH

LOW EARTH ORBIT

IRIDIUM SATELLITE

POLAR ORBIT

MAPPING SATELLITE

ACCORDING TO A RECENT COUNT, THERE ARE **129 MILLION OBJECTS** LARGER THAN **0.04 IN (1 MM)** IN ORBIT **AROUND THE EARTH**

Solar panels generate electricity to power satellite

Position of satellite controlled by stationary plasma thruster

2 Incoming signal amplified
Satellites boost the original radio signal using power from their solar panels. Onboard technology may be capable of processing many separate signals at once.

Fuel for thrusters stored in pressurized liquid propellant tanks

Anatomy of a comsat
Communications satellites (comsats) feature extremely sophisticated equipment designed to cope for extended periods of time in the extreme conditions of space, where maintenance is practically impossible. Power is generated by solar panels.

Reflector receives incoming radio signals and redirects them to antenna feed

Communications satellites

Many satellites act as relays for radio signals used in various types of communication. A satellite high above Earth can maintain a direct line of sight to receivers and transmitters on the ground below, allowing access to communications such as telephone, internet services, and satellite television even in remote areas beyond the range of ground-based radio transmitters. Satellites in geostationary orbit 22,236 miles (35,786 km) above Earth can remain stationary above a fixed point on the equator, hanging in the sky and acting as broadcast platforms for signals that can be picked up by receivers across a large expanse of Earth's surface.

Optical solar reflectors control satellite's temperature

Telemetry, tracking, and command antenna allows ground station to monitor and control satellite operations

Incoming radio signals are fed by antenna to transponder for processing; antenna sends outgoing signals back to Earth via reflector

RADIO SIGNALS

3 Signal transmitted back to Earth
The satellite retransmits the signal to Earth, either as a narrow beam directed to another ground station or as a broadcast signal that is weaker and more widely spread.

WHO INVENTED THE COMSAT?

The idea of a communications relay in geostationary orbit was proposed by science fiction author Arthur C. Clarke in 1948—although he thought such a relay would have to be a crewed space station.

1 Signal transmitted
Radio signals may be sent to the satellite from a ground station equipped with a powerful, directional dish antenna or from much weaker sources, such as the antenna on a satellite phone.

CUBESATS

While geostationary comsats must be large in order to generate enough power for relaying and broadcasting signals over long distances, sending signals to and from low Earth orbit (LEO) takes much less energy. Earth is now orbited by flocks of small comsats in LEO, often designed around an efficient, modular, and lightweight template called the cubesat.

Each cubesat unit is a 4-in (10-cm) cube

Multiple units with specialized functions locked together

1 UNIT

24 UNITS

4 Signal received
The receiver may either decode the radio signal, channel it into a ground-based communications network, or retransmit it to another comsat for relaying further around the world.

GROUND STATION

Types of satellite

Satellites have a wide variety of uses, but the vast majority are involved in communications and navigation, with applications ranging from steering supertankers to broadcasting television.

GPS and navigation satellites

Because radio signals travel at a known speed (the speed of light), it is possible to use time signals received from satellites in well-defined orbits to pin down a receiver's location on Earth. This is the basis of satellite navigation systems such as the Global Positioning System (GPS), which have become an indispensable part of modern technologies ranging from smartphones and cars to crop management.

Satellite 1
A timed signal from a single satellite locates a receiver at a known distance, somewhere on a spherical surface.

Receiver's distance from Satellite 1 is a point on a circle

EARTH

Satellite 2
Comparison with a signal from a second satellite reduces the possible location to two intersection points.

Location narrowed down to either of two points

Satellite 3
A third satellite signal will provide a single intersection point at sea level on Earth's surface.

Receiver location can now only be a single point

Satellite 4
A fourth satellite signal takes account of varying altitudes and provides a position in three dimensions.

Position confirmed to within 3 ft (1 m)

THE EUROPEAN **GALILEO** SATELLITE NAVIGATION SYSTEM CAN **PINPOINT** **POSITIONS** ON EARTH TO WITHIN **8 IN (20 CM) OR BETTER**

Looking back at Earth

Large numbers of satellites now monitor Earth's land surface, atmosphere, and oceans from space using a variety of techniques known as remote sensing.

Earth in many wavelengths

The idea of remote sensing began in the 1960s, when astronauts reported seeing surprising levels of detail from orbit. The first attempts at studying Earth from space involved simple photography, sometimes enhanced by telescopes. Since then, more advanced tools have been introduced, such as photographing the surface through filters to determine its response to light at specific wavelengths—a technique called multispectral imaging.

Analyzing crop health
Images of the ground taken at varied wavelengths of visible light and invisible heat radiation can reveal different properties and build up a picture of crop health for use by farmers.

MULTISPECTRAL IMAGING

Small amount of blue and red light returns; most absorbed to fuel photosynthesis

Lots of infrared light returned by healthy leaf

Less infrared light being reflected by stressed leaf

Less infrared and green light reflected by dead leaf

HEALTHY LEAF

STRESSED LEAF

DEAD LEAF

Multispectral imaging of crops works because leaves and other vegetation contain pigments that absorb certain wavelengths of light and reflect others. The health of the plant creates subtle changes in absorption and reflection that can be detected by measuring output at specific wavelengths.

Orbiting satellite

Sunlight illuminates crop

Reflected radiation detected by satellite

Pixels in satellite image correspond to areas on ground; the smaller the area, the higher the image resolution

OVERALL CROP HEALTH

Nitrogen levels higher in healthy plants

NITROGEN ABSORPTION LEVELS

Areas of drier crop shown in red

DRY BIOMASS LEVELS

Areas in need of fertilizer spraying

FERTILIZER LEVELS

FARMLAND

Weather satellites

Weather monitoring was one of the first applications of satellites. Photographing the atmosphere from high orbit allows a more detailed understanding of large-scale weather patterns, while radar systems study the effects of Earth's atmosphere and ocean surface on reflected radio beams in order to measure wind speed, rainfall, and wave heights. Satellites can also detect the levels of pollutants in Earth's atmosphere and measure the temperature to keep track of climate change.

Satellite train
The A-Train is a group of remote-sensing satellites in near-identical Sun-synchronous orbits that allow them to conduct daily, almost simultaneous observations of several atmospheric properties.

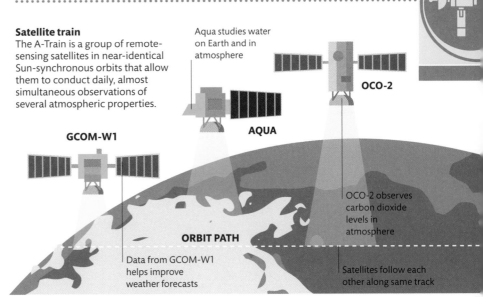

Aqua studies water on Earth and in atmosphere

OCO-2

AQUA

GCOM-W1

OCO-2 observes carbon dioxide levels in atmosphere

ORBIT PATH

Data from GCOM-W1 helps improve weather forecasts

Satellites follow each other along same track

REMOTE-SENSING TECHNOLOGIES
Satellites carry a wide variety of different tools and sensors, including spectrometers that analyze the absorption and reflection of light at different wavelengths and radar that can map Earth's landscape and oceans.

Meteorology
Photography of cloud patterns can be supplemented by radar measures of wind speed and rainfall and by infrared cameras that measure surface temperatures.

Oceanography
Radar instruments measure the speed and height of waves, revealing circulation patterns and wind speeds at sea. Infrared detectors can track ocean temperatures.

Geology
Hyperspectral imaging measures the complete spectrum of light reflected from Earth's surface. This can help identify specific rocks and minerals.

Surveying
Satellite-based radar can produce maps of terrain across large areas of the globe, while stereo photography of small areas can be used to create 3D models.

Land use
Multispectral imaging can help distinguish between areas of natural forest, agriculture, urban development, and water, revealing patterns of land use.

Archaeology
Satellite images and ground-penetrating radar can reveal the outlines and remains of ancient settlements and structures that have become buried over centuries.

IN 2011, 17 PREVIOUSLY UNKNOWN EGYPTIAN PYRAMIDS WERE UNCOVERED USING SATELLITE IMAGERY

ACTIVE AND PASSIVE REMOTE SENSING

Remote sensing systems that measure naturally available energy are called passive sensors. Passive remote-sensing instruments can only be used to detect energy when it is naturally available. Active remote-sensing instruments can fire out signals using their own energy source and analyze the results.

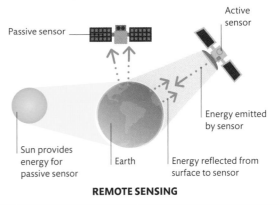

Passive sensor

Active sensor

Energy emitted by sensor

Sun provides energy for passive sensor

Earth

Energy reflected from surface to sensor

REMOTE SENSING

Looking farther into space

Satellite-based astronomical observatories can study the Universe in new ways, capturing perfect images free from turbulence and detecting radiation that is blocked by Earth's atmosphere.

Space telescope orbits

While standard low Earth orbit is sufficient for many space telescopes, some missions require more complex orbits. More distant orbits reduce the apparent size of Earth and make more of the sky visible at any one time, while some satellites follow Earth-trailing orbits around the Sun in order to avoid their instruments being swamped by Earth's radiation. Placing satellites in special locations called Lagrangian points ensures that Earth and the Sun remain fixed in the same orientation relative to the satellite.

WHICH IS THE BIGGEST SPACE TELESCOPE?

Planned for launch in 2021, NASA's giant James Webb Space Telescope has a 21-ft (6.4-m) mirror. It will orbit at the Earth–Sun L1 point, four times farther from Earth than the Moon.

150,000
DISTANT STARS WERE SIMULTANEOUSLY MONITORED BY THE KEPLER SATELLITE

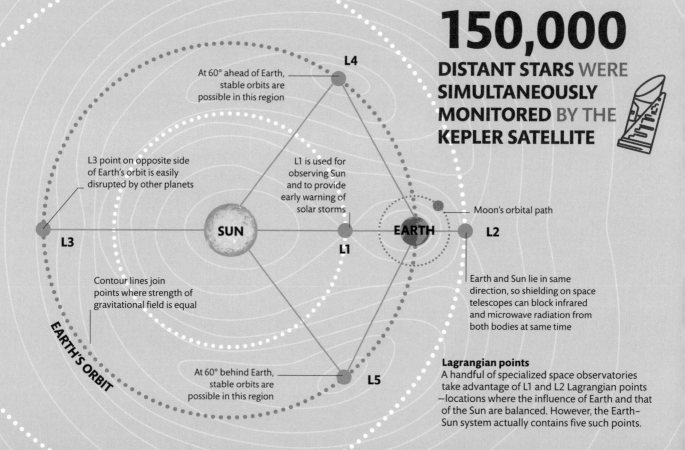

- **L4** — At 60° ahead of Earth, stable orbits are possible in this region
- **L3** point on opposite side of Earth's orbit is easily disrupted by other planets
- **L1** is used for observing Sun and to provide early warning of solar storms
- Moon's orbital path
- **L3**
- **SUN**
- **EARTH**
- **L2**
- **L1**
- Contour lines join points where strength of gravitational field is equal
- Earth and Sun lie in same direction, so shielding on space telescopes can block infrared and microwave radiation from both bodies at same time
- **EARTH'S ORBIT**
- At 60° behind Earth, stable orbits are possible in this region
- **L5**

Lagrangian points

A handful of specialized space observatories take advantage of L1 and L2 Lagrangian points —locations where the influence of Earth and that of the Sun are balanced. However, the Earth–Sun system actually contains five such points.

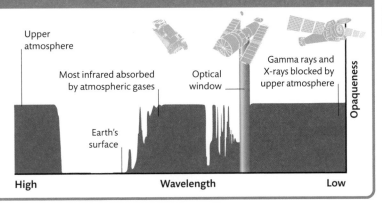

DETECTING BLOCKED RADIATION

One major advantage of space-based astronomy is the ability to detect radiation that is blocked by Earth's atmosphere. High-energy electromagnetic rays beyond the near-ultraviolet are entirely absorbed by the atmosphere (fortunately for life), while at the other end of the spectrum, much infrared radiation and many longer radio waves are all absorbed. Warm water vapor in the lower atmosphere also releases infrared radiation that can swamp the weak rays from space.

Upper atmosphere

Most infrared absorbed by atmospheric gases

Optical window

Gamma rays and X-rays blocked by upper atmosphere

Earth's surface

Opaqueness

High · Wavelength · Low

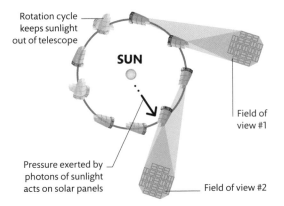

Rotation cycle keeps sunlight out of telescope

SUN

Field of view #1

Pressure exerted by photons of sunlight acts on solar panels

Field of view #2

Looking for planets

NASA's Kepler Space Telescope was a satellite launched in 2009 to detect alien planets by measuring minute dips in starlight as they pass in front of their parent stars. Placed in an Earth-trailing orbit, its initial mission involved keeping an unblinking eye on a crowded cloud of stars in the constellation Cygnus, which it did for more than three years from 2009.

The Kepler mission
Following failures in Kepler's pointing technology in 2013, engineers found an ingenious way to stabilize it using pressure from sunlight, allowing it to continue studying different parts of the sky for shorter periods.

High-energy astronomy

High-energy astronomy satellites image the Universe using ultraviolet (UV) radiation, X-rays, and gamma rays that are produced by some of the hottest and most violent objects in space but cannot be detected at Earth's surface. While UV can be focused using traditional telescope designs, the energy of X-rays and gamma rays allows them to pass through normal mirrors, so other designs must be used.

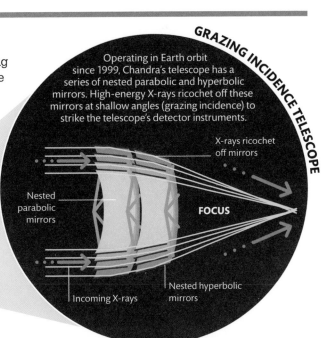

GRAZING INCIDENCE TELESCOPE

Operating in Earth orbit since 1999, Chandra's telescope has a series of nested parabolic and hyperbolic mirrors. High-energy X-rays ricochet off these mirrors at shallow angles (grazing incidence) to strike the telescope's detector instruments.

X-rays ricochet off mirrors

Nested parabolic mirrors

FOCUS

Incoming X-rays

Nested hyperbolic mirrors

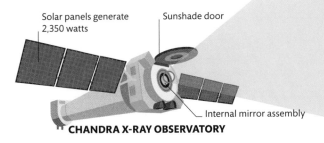

Solar panels generate 2,350 watts

Sunshade door

Internal mirror assembly

CHANDRA X-RAY OBSERVATORY

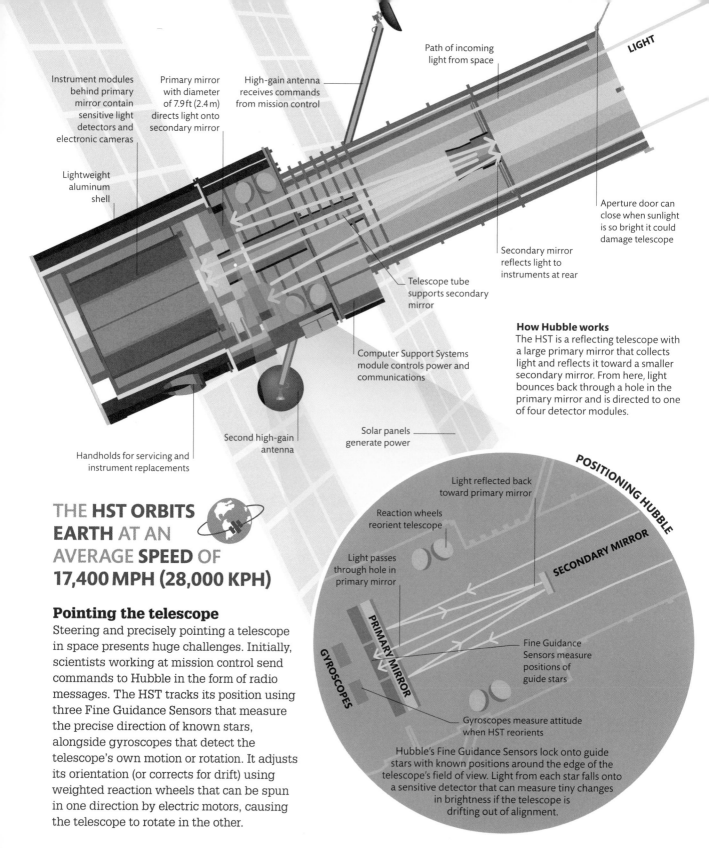

Instrument modules behind primary mirror contain sensitive light detectors and electronic cameras

Primary mirror with diameter of 7.9 ft (2.4 m) directs light onto secondary mirror

High-gain antenna receives commands from mission control

Path of incoming light from space

LIGHT

Lightweight aluminum shell

Aperture door can close when sunlight is so bright it could damage telescope

Secondary mirror reflects light to instruments at rear

Telescope tube supports secondary mirror

Computer Support Systems module controls power and communications

How Hubble works
The HST is a reflecting telescope with a large primary mirror that collects light and reflects it toward a smaller secondary mirror. From here, light bounces back through a hole in the primary mirror and is directed to one of four detector modules.

Handholds for servicing and instrument replacements

Second high-gain antenna

Solar panels generate power

THE HST ORBITS EARTH AT AN AVERAGE SPEED OF 17,400 MPH (28,000 KPH)

Pointing the telescope
Steering and precisely pointing a telescope in space presents huge challenges. Initially, scientists working at mission control send commands to Hubble in the form of radio messages. The HST tracks its position using three Fine Guidance Sensors that measure the precise direction of known stars, alongside gyroscopes that detect the telescope's own motion or rotation. It adjusts its orientation (or corrects for drift) using weighted reaction wheels that can be spun in one direction by electric motors, causing the telescope to rotate in the other.

POSITIONING HUBBLE

Light reflected back toward primary mirror

Reaction wheels reorient telescope

SECONDARY MIRROR

Light passes through hole in primary mirror

PRIMARY MIRROR

GYROSCOPES

Fine Guidance Sensors measure positions of guide stars

Gyroscopes measure attitude when HST reorients

Hubble's Fine Guidance Sensors lock onto guide stars with known positions around the edge of the telescope's field of view. Light from each star falls onto a sensitive detector that can measure tiny changes in brightness if the telescope is drifting out of alignment.

The Hubble Space Telescope

The Hubble Space Telescope (HST) is the largest
and most successful space telescope (see pp.22–23),
operating in Earth orbit for more than 30 years
and producing thousands of discoveries that have
revolutionized our understanding of the Universe.

HOW MANY TIMES HAS HUBBLE BEEN SERVICED?

Since its launch in 1990, the
HST has been repaired and
upgraded in space on five
separate missions—most
recently in 2009, shortly
before the Space Shuttle
was retired.

What Hubble sees

From its location in low Earth
orbit, the HST can produce images
whose detail is limited only by the
dimensions of its mirror and the
sensitivity of its instruments. In
practice, this means that, although
the telescope is relatively modest
by today's standards, its pictures
can rival those from much larger
Earth-based observatories
(see pp.24–25). Furthermore, the
lack of atmospheric absorption
means that some of the HST's
instruments can detect invisible
radiation from the near-infrared
to the near-ultraviolet, revealing
material too cool or too hot to shine
in visible light.

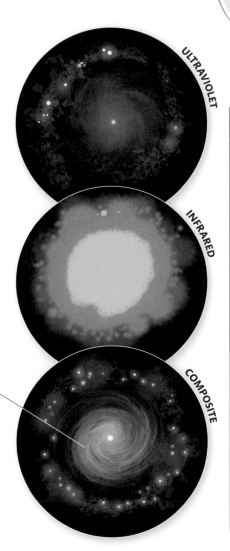

ULTRAVIOLET

INFRARED

COMPOSITE

Composite image of spiral
galaxy NGC 1512, 38 million
light-years away from Earth

Wavelengths

Combining near-infrared maps of relatively
cool cosmic dust with ultraviolet views of a
galaxy's hottest stars, the Hubble Space
Telescope can build up a complete picture
of structures situated in a distant galaxy.

MANAGING THE DATA

Data from the various HST
instruments is initially stored on
the telescope itself. Every 12 hours
or so, it is uploaded to one of
NASA's Tracking and Data Relay
Satellites in high geostationary
orbit, from where it is relayed to a
ground station in New Mexico.
From here, it is passed to the HST
control center in Maryland and on
to the Space Telescope Science
Institute in Baltimore.

LIGHT → HST → SATELLITE

SCIENCE INSTITUTE ← GROUND STATION

Anatomy of a space probe

A probe is a small, uncrewed spacecraft carrying scientific instruments that gather data about the environment of space and distant objects the probe visits. The instruments may detect particles, measure electrical and magnetic fields, and produce images of objects. The probe also carries subsystems that allow it to operate in space and carry out its job. These include engines for changing the probe's orientation and orbit, radio equipment to receive instructions from Earth and send back scientific data, computers to control its operations, and power systems and environmental controls to keep all systems running.

SUN

Intense electric and magnetic fields

Hot gas outbursts from Sun

High-energy particles from solar flares

Solar wind of particles from Sun's upper atmosphere

1 **Gathering data**
The probe is continually bombarded by fierce radiation and energetic particles—its design shields it from damaging effects while allowing it to measure conditions and detect particles.

Temperatures on heat shield reach up to 2,500°F (1,370°C)

Probing the Sun
The Parker Solar Probe is a spacecraft designed to fly through the harsh environment close to the Sun, measuring magnetic fields and collecting the high-energy particles that the Sun ejects.

Heat shield protects sensitive instruments

Antennae measure electric fields

Solar array cooling system

Solar panels generate energy and cool spacecraft

Particle detector registers solar winds

Magnetometer measures magnetic fields

Probe comes within 12 million miles (19 million km) of Sun

2 **Communication with Earth**
Data from five different scientific instruments is processed by the onboard computer and converted into electric signals. A small dish-shaped antenna sends the data to Earth via high-frequency radio waves.

Space probes and orbiters

Space probes are robot spacecraft that enter another planet's atmosphere or land on the surface of another body to gather scientific data. Orbiters are not designed to penetrate the atmospheres of other bodies.

Parabolic dish collects and focuses radio waves

RADIO TELESCOPE

Antenna creates electric current

3 **Receiving signals**
Large radio dishes on Earth receive the probe's signals. The dish focuses waves gathered across a large area onto a small receiver, which generates a weak current.

HOW LONG WOULD IT TAKE TO SEND A SPACECRAFT TO THE STARS?

Traveling at 38,000 mph (61,000 kph), the Voyager 1 spacecraft is the fastest object leaving the Solar System, but it would take 70,000 years to reach the nearest star.

THE FASTEST SPACE PROBE EVER LAUNCHED, THE PARKER SOLAR PROBE, ACHIEVED A SPEED OF 244,000 MPH (393,000 KPH)

5 **Decoding the data**
Scientists use computers that decode the raw numbers into useful data and process it to make images, graphs, and other "data products."

Data decoded and processed by computers

COMPUTER

Data sent to laboratory

RECEIVER AND AMPLIFIER

Current flows to receiver

4 **Amplification**
An amplifier takes the raw signal, boosts its strength, and decodes it into digital data (pulses that represent the strength of signals gathered by the probe).

Reaching other worlds

In order to reach distant planets or other objects, a probe must first reach escape velocity to break free of Earth's gravity before entering a transfer orbit around the Sun (see p.181). The shape of this orbit (or a segment of it) bridges the gap to where the target object will be at a future point in time, where the spacecraft can then slow down and allow itself to be captured by its target's gravity. The different orbital speeds of objects at different distances from the Sun add to the complications.

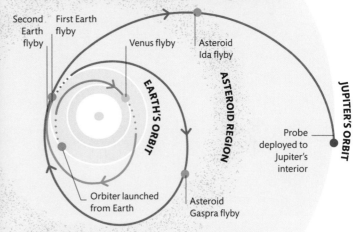

Second Earth flyby
First Earth flyby
Venus flyby
Asteroid Ida flyby
EARTH'S ORBIT
ASTEROID REGION
JUPITER'S ORBIT
Probe deployed to Jupiter's interior
Orbiter launched from Earth
Asteroid Gaspra flyby

Galileo's flight trajectory
The Galileo orbiter's five-year journey to Jupiter involved two flybys of Earth and one of Venus. The orbiter altered its trajectory and gained speed on each flyby.

HEAT SHIELDING

Probes exploring the inner Solar System require thick shielding to protect instruments from the scorching heat on their sunlit sides. The design must also distribute heat to avoid stress between hot and cold parts of the spacecraft.

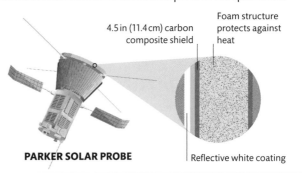

4.5 in (11.4 cm) carbon composite shield
Foam structure protects against heat

PARKER SOLAR PROBE
Reflective white coating

Propulsion in space

While chemical rockets are necessary to lift spacecraft away from Earth's surface, several more efficient forms of propulsion can be used in orbit and beyond.

Electrons and xenon atoms collide

IONIZATION CHAMBER

How an ion engine works

An ion thruster transforms neutral atoms of a gas (usually xenon) into electrically charged ions. It then accelerates them to high speed in a high-voltage electric field, expelling them into space to generate thrust.

POSITIVELY CHARGED GRID

NEGATIVELY CHARGED GRID

EQUAL BUT OPPOSITE FORCES

KEY

Xenon

Xenon ion

Electron

Xenon ions escape thruster

High voltage between grids accelerates xenon ions

4 Expelled ions
Ions escape from the rear of the thruster, creating a small thrust force with high efficiency. The spacecraft is pushed forward by an equal but opposite force.

3 Acceleration
Xenon ions are accelerated to a high speed by an intense electric field generated by the voltage between two oppositely charged electrode grids.

Ion engines

Ion thrusters generate a small amount of thrust by expelling electrically charged particles (ions) at extremely high speeds. This allows the engine to run for months with the potential to reach high speeds and cover great distances, expending only tiny amounts of fuel. Ion engines have been used in several spacecraft, including the Dawn mission to the asteroids Ceres and Vesta (see pp.62–63).

THE **THRUST** PRODUCED BY **DAWN'S ION ENGINE** IS EQUIVALENT TO THE **WEIGHT OF TWO SHEETS OF A4 PAPER** RESTING ON YOUR HAND

HOW LONG CAN AN ION ENGINE RUN FOR?

During an 11-year mission, NASA's Dawn spacecraft ran its ion engine for a total of 5.9 years, altering its speed by a total of 5,700 miles (41,400 km) per hour.

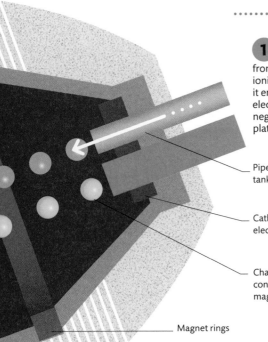

1 **Propellant released**
Xenon is injected from storage tanks into an ionization chamber, where it encounters fast-moving electrons emitted by a hot, negatively charged magnetic plate known as a cathode.

Pipe from propellant tank injects xenon

Cathode heated by electricity from solar cells

Charged particles confined by magnetic field

Magnet rings

2 **Creating ions**
Electrons collide with xenon atoms, stripping away electrons from the propellant's outer layers and transforming them into positively charged ions.

Maneuvering in space

Many spacecraft and satellites are equipped with thrusters that fire small jets of gas to push themselves around and change their orientation. Fuel is a precious commodity in space, so maneuvers must be meticulously planned. For precise alignments, some spacecraft use reaction wheels—motorized disks that can spin around one axis, causing the spacecraft body to rotate the opposite way.

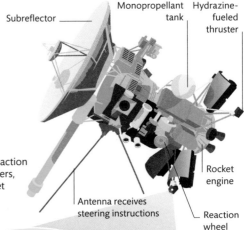

Subreflector

Monopropellant tank

Hydrazine-fueled thruster

Rocket engine

Antenna receives steering instructions

Reaction wheel

Orientation in space
A spacecraft like NASA's Cassini orbiter uses a combination of reaction wheels, hydrazine-fueled thrusters, and a traditional chemical rocket engine to adjust its orientation.

SOLAR SAILS

Solar sails harness the pressure exerted by light streaming out from the Sun. Despite lacking mass, photons of light carry momentum that can transfer to a large reflecting surface. Solar sails, like ion engines, produce tiny amounts of thrust for extremely long periods. The technology was first successfully tested in Japan's IKAROS spacecraft in 2010.

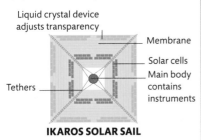

Liquid crystal device adjusts transparency

Membrane

Solar cells

Main body contains instruments

Tethers

IKAROS SOLAR SAIL

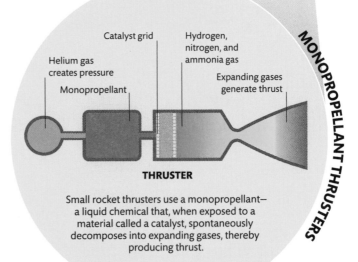

Helium gas creates pressure

Catalyst grid

Hydrogen, nitrogen, and ammonia gas

Expanding gases generate thrust

Monopropellant

THRUSTER

MONOPROPELLANT THRUSTERS

Small rocket thrusters use a monopropellant—a liquid chemical that, when exposed to a material called a catalyst, spontaneously decomposes into expanding gases, thereby producing thrust.

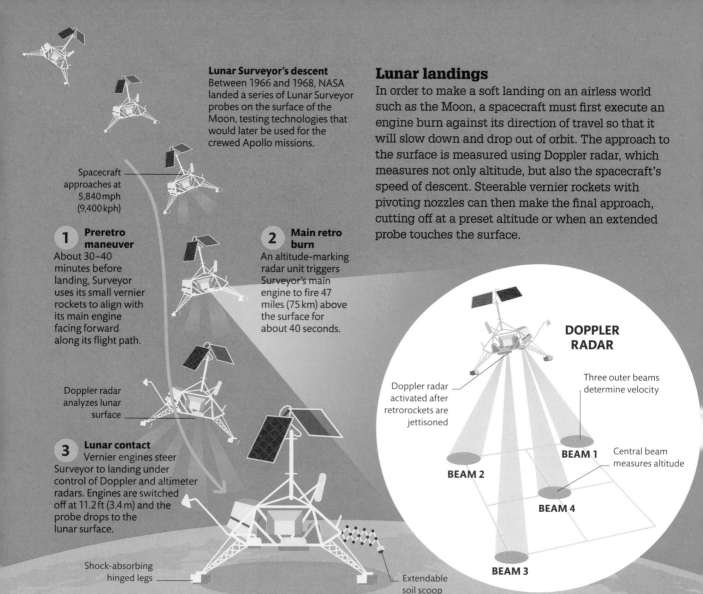

Lunar Surveyor's descent
Between 1966 and 1968, NASA landed a series of Lunar Surveyor probes on the surface of the Moon, testing technologies that would later be used for the crewed Apollo missions.

Lunar landings

In order to make a soft landing on an airless world such as the Moon, a spacecraft must first execute an engine burn against its direction of travel so that it will slow down and drop out of orbit. The approach to the surface is measured using Doppler radar, which measures not only altitude, but also the spacecraft's speed of descent. Steerable vernier rockets with pivoting nozzles can then make the final approach, cutting off at a preset altitude or when an extended probe touches the surface.

Spacecraft approaches at 5,840 mph (9,400 kph)

1   Preretro maneuver
About 30–40 minutes before landing, Surveyor uses its small vernier rockets to align with its main engine facing forward along its flight path.

2   Main retro burn
An altitude-marking radar unit triggers Surveyor's main engine to fire 47 miles (75 km) above the surface for about 40 seconds.

Doppler radar analyzes lunar surface

3   Lunar contact
Vernier engines steer Surveyor to landing under control of Doppler and altimeter radars. Engines are switched off at 11.2 ft (3.4 m) and the probe drops to the lunar surface.

Shock-absorbing hinged legs

Extendable soil scoop

DOPPLER RADAR

Doppler radar activated after retrorockets are jettisoned

Three outer beams determine velocity

BEAM 1

Central beam measures altitude

BEAM 2

BEAM 4

BEAM 3

Soft landings

Landing on airless worlds is a relatively simple, though delicate, task. With no air resistance to reduce its speed, a spacecraft must slow its descent to the surface through the use of rockets.

ROSETTA LANDED ON COMET 67P **AT A SPEED OF** LESS THAN **3 FT (1 M) PER SECOND**

WHAT WAS THE FIRST SOFT LANDING ON ANOTHER WORLD?

The first space probe to make a soft landing was the Soviet Union's Luna 9. It used airbags to survive a 14-mph (22-kph) impact on the Moon in 1965.

Drifting to touchdown

Spacecraft orbiting around low-gravity bodies such as comets and asteroids can simply adjust their orbits through a series of short engine burns from their thrusters. These spacecraft gradually spiral inward in order to deliver more detailed views of the target object and eventually make a gentle touchdown on the object's surface.

8,200 FT (2,500 M)

PHILAE

6,600 FT (2,000 M)

4,900 FT (1,500 M)

3,300 FT (1,000 M)

1,600 FT (500 M)

Landing on a comet

Following Rosetta's arrival at Comet 67P in 2014, the spacecraft released a small lander called Philae. Unlike the main spacecraft, this was deliberately designed to touch down on the comet's surface, taking photos as it descended.

1 Philae alone
The Philae probe separated from Rosetta at an altitude of 12 miles (20 km), with a release mechanism that pushed it onto a descent path toward the comet.

First bounce reaches altitude of 0.6 miles (1 km)

2 Touchdown
On contact with the surface, a gas thruster on the rear of the probe was intended to fire, in order to push it onto the comet and avoid bouncing, before two harpoons anchored it to the rocky surface.

Rosetta's trajectory

At the end of its mission to Comet 67P in September 2016, the European Space Agency's Rosetta probe was steered to a gentle crash-landing on the surface.

1 Final orbits
Rosetta's last complete orbits around the comet came to within 3 miles (5 km) of its surface.

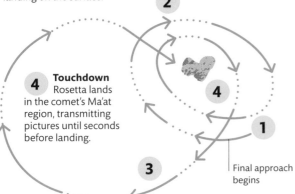

4 Touchdown
Rosetta lands in the comet's Ma'at region, transmitting pictures until seconds before landing.

Final approach begins

3 One last burn
A final 208-second engine burn at 12 miles (19 km) puts the probe on a straight descent path toward its landing site.

2 Outward swing
After a second orbit of the comet, Rosetta's course is corrected in preparation for descent and landing.

LANDING 1

LANDING 2

LANDING 3

3 Bouncing twice
Later analysis showed that Philae's harpoons did not trigger. Instead, the lander bounced off the surface twice before coming to rest on its side in a shadowed crevice where its solar panels were unable to recharge.

CRASH LANDINGS

Sometimes spacecraft are deliberately crashed onto a planetary surface at high speed. NASA's Deep Impact probe carried a barrel-shaped projectile that smashed into the surface of comet Tempel 1 in 2005 so the main spacecraft could study the debris thrown up.

TEMPEL 1

Deep Impact's projectile

Plume of ejected debris analyzed by main spacecraft

Crewed spacecraft

Spacecraft that carry astronauts have to be both larger and more complex than robot probes because they must carry specialized equipment to keep the astronauts alive and protect them during reentry.

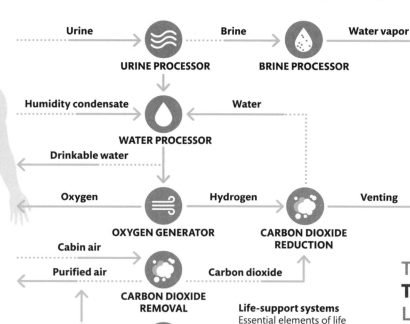

Urine → **URINE PROCESSOR** → Brine → **BRINE PROCESSOR** → Water vapor

Humidity condensate → **WATER PROCESSOR** ← Water

Drinkable water ← **WATER PROCESSOR**

Oxygen ← **OXYGEN GENERATOR** → Hydrogen → **CARBON DIOXIDE REDUCTION** → Venting

Cabin air → **CARBON DIOXIDE REMOVAL**

Purified air ← **CARBON DIOXIDE REMOVAL** ← Carbon dioxide

Cabin air → **TRACE CONTAMINANT CONTROL**

Life-support systems
Essential elements of life support include providing drinkable water and breathable oxygen (usually extracted from water), removing toxic carbon dioxide, and processing waste.

Aluminum-alloy hull

Multiple layers of thermal and impact insulation

Radio antenna for docking

Docking assembly probe

ORBITAL MODULE

Cabin filled with Earth-like nitrogen and oxygen atmosphere at normal surface pressure

THERE HAVE BEEN **MORE THAN 140 SUCCESSFUL** LAUNCHES OF **SOYUZ**

CREWED SPACECRAFT VEHICLES

Since the first Russian and US astronauts flew into space in 1961, there have been well over 300 successful crewed spaceflights. Although men and women of many nationalities have now become astronauts, only three countries—the United States, the Soviet Union (modern-day Russia), and China—have developed and launched their own crewed spacecraft.

	SOYUZ	**APOLLO**	**SHENZHOU**	**ORION**
Country	Russia	US	China	US
Crew	3	3	3	4–6
Operational	1967–present	1968–1975	2003–present	2023–
Length	24.5 ft (7.5 m)	36 ft (11 m)	30 ft (9 m)	26 ft (8 m)

SPLASHDOWN

For spacecraft aiming to land in the ocean, a swift recovery is key. In 2020, SpaceX's Crew Dragon Demo-2 mission completed the first splashdown in 45 years, landing within sight of waiting recovery boats.

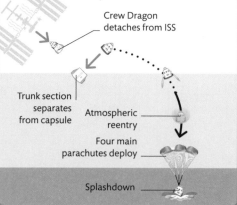

Crew Dragon detaches from ISS

Trunk section separates from capsule

Atmospheric reentry

Four main parachutes deploy

Splashdown

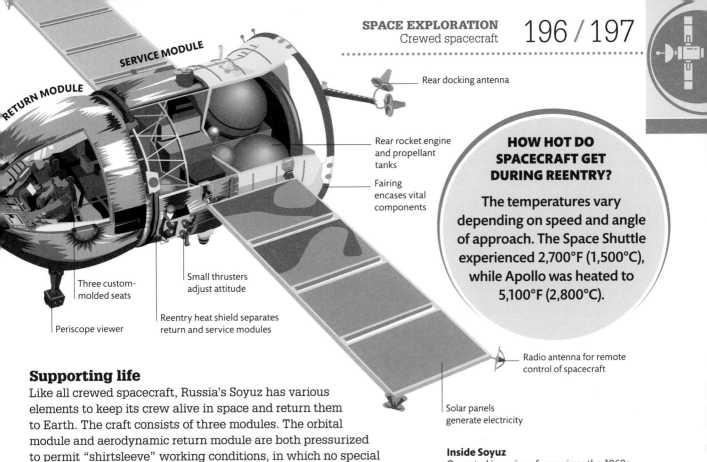

SERVICE MODULE

RETURN MODULE

Rear docking antenna

Rear rocket engine
and propellant
tanks

Fairing
encases vital
components

Three custom-
molded seats

Small thrusters
adjust attitude

Periscope viewer

Reentry heat shield separates
return and service modules

Radio antenna for remote
control of spacecraft

Solar panels
generate electricity

HOW HOT DO SPACECRAFT GET DURING REENTRY?

The temperatures vary depending on speed and angle of approach. The Space Shuttle experienced 2,700°F (1,500°C), while Apollo was heated to 5,100°F (2,800°C).

Supporting life

Like all crewed spacecraft, Russia's Soyuz has various elements to keep its crew alive in space and return them to Earth. The craft consists of three modules. The orbital module and aerodynamic return module are both pressurized to permit "shirtsleeve" working conditions, in which no special clothing need be worn. An unpressurized service module provides power, propulsion, and supplies for the life-support systems.

Inside Soyuz
Operated in various forms since the 1960s, Soyuz can support up to three crew members and is capable of docking with other spacecraft.

Returning to Earth

Most returning spacecraft rely on friction with the air during reentry to slow their descent to a point where parachutes can open. The reentry or descent module is fitted with a heat shield designed to ablate (break away, carrying heat with it), and its design is usually conical to ensure that the spacecraft aligns itself to bear the brunt of the heating on its wide base. US spacecraft have traditionally splashed down in the ocean with recovery ships standing by, while Russian and Chinese capsules returning over land use retrorockets to slow their final descent.

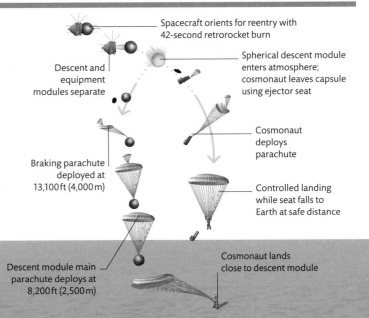

Spacecraft orients for reentry with 42-second retrorocket burn

Descent and equipment modules separate

Spherical descent module enters atmosphere; cosmonaut leaves capsule using ejector seat

Cosmonaut deploys parachute

Braking parachute deployed at 13,100 ft (4,000 m)

Controlled landing while seat falls to Earth at safe distance

Descent module main parachute deploys at 8,200 ft (2,500 m)

Cosmonaut lands close to descent module

Safe landing
Cosmonauts on early Soviet spaceflights, such as Vostok 1, ejected from their spacecraft after reentry and parachuted separately to Earth for a safe return. From 1964, Voskhod missions saw cosmonauts land in the reentry capsule.

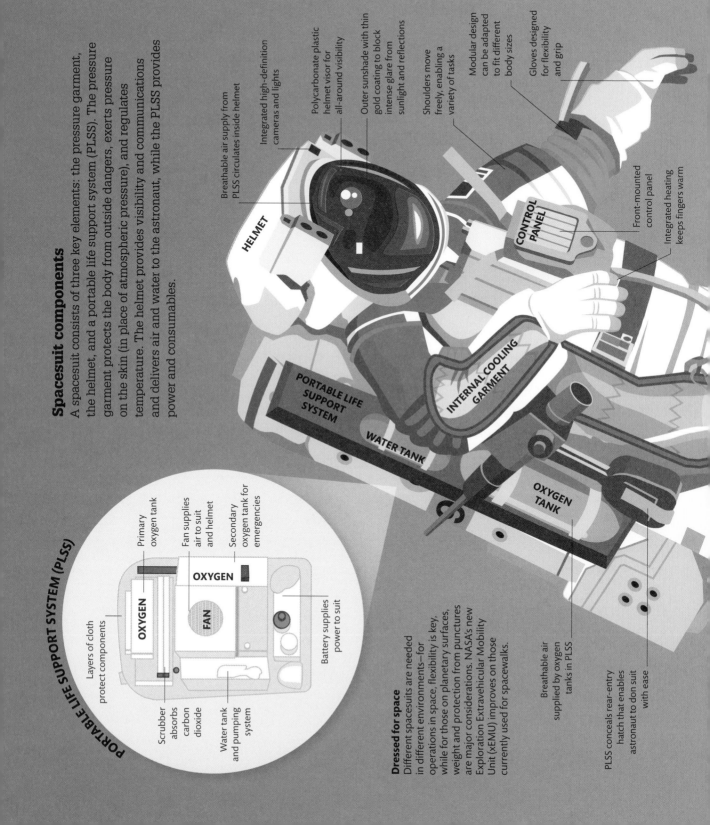

Spacesuit components

A spacesuit consists of three key elements: the pressure garment, the helmet, and a portable life support system (PLSS). The pressure garment protects the body from outside dangers, exerts pressure on the skin (in place of atmospheric pressure), and regulates temperature. The helmet provides visibility and communications and delivers air and water to the astronaut, while the PLSS provides power and consumables.

Breathable air supply from PLSS circulates inside helmet

Integrated high-definition cameras and lights

Polycarbonate plastic helmet visor for all-around visibility

Outer sunshade with thin gold coating to block intense glare from sunlight and reflections

Shoulders move freely, enabling a variety of tasks

Modular design can be adapted to fit different body sizes

Gloves designed for flexibility and grip

HELMET

CONTROL PANEL

Front-mounted control panel

Integrated heating keeps fingers warm

PORTABLE LIFE SUPPORT SYSTEM

INTERNAL COOLING GARMENT

WATER TANK

OXYGEN TANK

PORTABLE LIFE SUPPORT SYSTEM (PLSS)

Layers of cloth protect components

Primary oxygen tank

Fan supplies air to suit and helmet

Secondary oxygen tank for emergencies

OXYGEN

OXYGEN

FAN

Battery supplies power to suit

Scrubber absorbs carbon dioxide

Water tank and pumping system

Dressed for space

Different spacesuits are needed in different environments—for operations in space, flexibility is key, while for those on planetary surfaces, weight and protection from punctures are major considerations. NASA's new Exploration Extravehicular Mobility Unit (xEMU) improves on those currently used for spacewalks.

Breathable air supplied by oxygen tanks in PLSS

PLSS conceals rear-entry hatch that enables astronaut to don suit with ease

Spacesuits

Spacesuits are complete self-contained environments designed to protect astronauts from hostile surroundings and provide the supplies they need while operating outside their spacecraft in the near-vacuum of empty space or on another world.

The dangers of radiation

Operating beyond Earth's atmosphere and outside a spacecraft, a spacesuit must offer some degree of protection from a variety of types of harmful radiation and particles.

Solar flares
High-energy particles from the Sun create electromagnetic problems that disrupt electronics.

Cosmic rays
Fast-moving particles and high-energy radiation from outside the Solar System pass through materials.

Ultraviolet radiation
Intense visible light and strong ultraviolet radiation can damage an astronaut's eyesight.

Trapped radiation
Particles in the Van Allen Belts around Earth can damage cells in an astronaut's body.

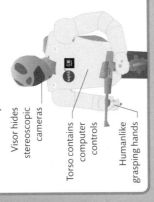

AN **ASTRONAUT** CAN **GROW** UP TO 3% TALLER WHILE LIVING IN SPACE

Three layers of stretchy spandex material maintain pressure on skin surface

Hiking-style boots with flexible soles for easy walking

SAFETY TETHER

Enhanced mobility at hips and knees for ease of movement in low gravity

Outer layers designed to resist shards of lunar dust and damage from micrometeoroids

FOOT RESTRAINTS

PIVOTING MOUNT

Pivoting mount with foot restraint and tether for working outside a spacecraft

ROBONAUT

To cut the number of extra-vehicular activities (EVAs) astronauts have to perform, NASA has developed its humanoid Robonaut to carry out routine tasks in and around the International Space Station.

Visor hides stereoscopic cameras

Torso contains computer controls

Humanlike grasping hands

WHO MADE THE FIRST SPACEWALK?

Russian cosmonaut Alexei Leonov became the first person to walk in space, leaving his Voskhod 2 spacecraft for 12 minutes and 9 seconds on March 18, 1965.

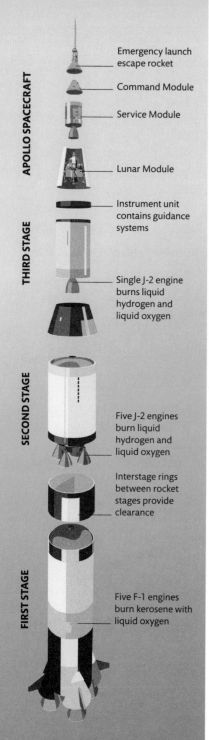

APOLLO SPACECRAFT
- Emergency launch escape rocket
- Command Module
- Service Module
- Lunar Module

THIRD STAGE
- Instrument unit contains guidance systems
- Single J-2 engine burns liquid hydrogen and liquid oxygen

SECOND STAGE
- Five J-2 engines burn liquid hydrogen and liquid oxygen
- Interstage rings between rocket stages provide clearance

FIRST STAGE
- Five F-1 engines burn kerosene with liquid oxygen

Launching Apollo
Sending Apollo to the Moon required a rocket with unprecedented power. Saturn V's three stages lifted it to Earth orbit, and once it broke free of Earth's gravity, the third stage reignited to put the spacecraft on a translunar flight path.

Mission to the Moon

Between 1969 and 1972, six US Apollo missions successfully carried astronauts to the Moon. Each expedition involved the launch of a complex three-part spacecraft using the enormous Saturn V rocket.

Apollo's journey
By using a separate Lunar Module for landing while keeping the larger CSM in orbit, the mass of payload launched from Earth was greatly reduced—at the cost of requiring complex and untried rendezvous operations.

7 Reentry
Approaching Earth, the Command Module (CM) separates from the Service Module (SM), turning 180° to reenter Earth's atmosphere.

Command Module rotates 180° to enter atmosphere with heat shield facing down

8 Splashdown
After reentry, parachutes deploy to slow the CM for splashdown in the Pacific Ocean. Flotation devices deploy while the crew await the arrival of recovery aircraft and boats.

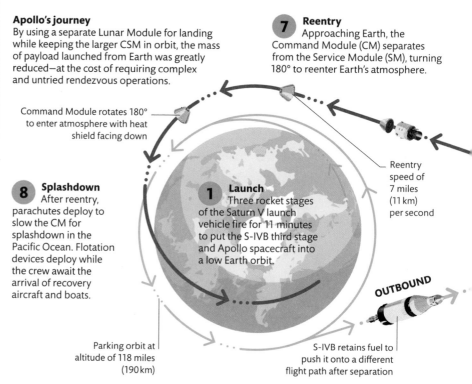

1 Launch
Three rocket stages of the Saturn V launch vehicle fire for 11 minutes to put the S-IVB third stage and Apollo spacecraft into a low Earth orbit.

Reentry speed of 7 miles (11 km) per second

OUTBOUND

Parking orbit at altitude of 118 miles (190 km)

S-IVB retains fuel to push it onto a different flight path after separation

To the Moon and back
Each Apollo mission involved sending three astronauts roughly 250,000 miles (400,000 km) to the Moon. One crew member remained in lunar orbit aboard the Command and Service Module (CSM), while the other two descended to the surface in the Lunar Module (LM). At the end of surface operations, the upper half of the LM blasted off to rendezvous with the CSM in lunar orbit for the return to Earth. Finally, the Command Module separated from the rest of the spacecraft for atmospheric reentry.

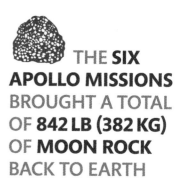

THE **SIX APOLLO MISSIONS** BROUGHT A TOTAL OF **842 LB (382 KG)** OF **MOON ROCK** BACK TO EARTH

Lunar lander

Designed to fly only in a near-vacuum, the ungainly Apollo Lunar Module consisted of a spiderlike Descent Stage and a pressurized Ascent Stage designed to carry two astronauts. Each stage had its own engine, allowing the Ascent Stage to return to lunar orbit at the end of its surface mission.

Landing on the lunar surface

The final stages of descent to the lunar surface involved precise piloting using the main descent engine and four reaction control thrusters—small multidirectional rockets positioned around the Ascent Stage.

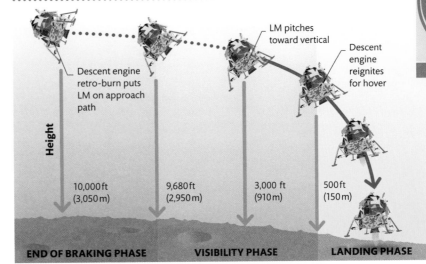

Descent engine retro-burn puts LM on approach path

LM pitches toward vertical

Descent engine reignites for hover

Height

10,000 ft (3,050 m)

9,680 ft (2,950 m)

3,000 ft (910 m)

500 ft (150 m)

END OF BRAKING PHASE

VISIBILITY PHASE

LANDING PHASE

3 **Lander attachment**
The CSM turns through 180° before docking with the LM Ascent Module and pulling it free from its housing.

CSM docked with LM Ascent Module

4 **Orbit and landing**
A CSM engine burn slows the spacecraft to put it into lunar orbit. Two astronauts board the LM and descend to the surface.

LM has been discarded

Final stage of Saturn V has been jettisoned

INBOUND

MOON

Lunar Module descent orbit insertion

5 **Lunar orbit rendezvous**
The LM Ascent Stage blasts off after the surface mission and docks with the CSM in lunar orbit. Astronauts and samples are transferred before the LM is discarded.

2 **Translunar injection**
After initial safety checks, the S-IVB rocket reignites to boost the spacecraft onto a translunar trajectory before separating and falling away.

6 **Return home**
The CSM fires engines to put the spacecraft on a return course toward Earth. The crossing between Earth and the Moon takes two to three days.

HOW MANY TESTS WERE PERFORMED BEFORE THE LANDING?

Only four crewed Apollo missions, numbered 7–10, flew before the Moon landing to test the spacecraft in Earth and lunar orbit.

THE LUNAR ROVING VEHICLE

The final three Apollo missions carried a Lunar Roving Vehicle that extended the range of exploration around the landing site. The lightweight but robust battery-powered vehicle could carry about twice its own weight and achieve a top speed of 11 mph (18 kph).

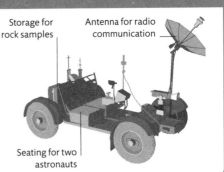

Storage for rock samples

Antenna for radio communication

Seating for two astronauts

The Shuttle at work

Once in space with its typical crew of around seven astronauts and payload specialists, the Shuttle orbiter was capable of carrying out a wide variety of different tasks. A large, pressurized cabin area provided living quarters, as well as room to house some experiments, while the huge, depressurized cargo bay could be used to carry experiments, deploy satellites and retrieve them from orbit for servicing, and deliver components for the International Space Station. The cargo bay could also carry a large, pressurized module called Spacelab to provide expanded laboratory space for experiments and servicing missions.

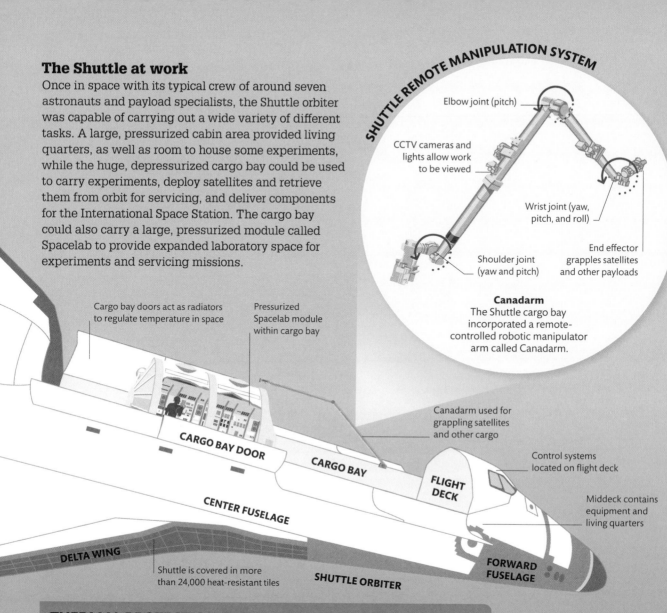

SHUTTLE REMOTE MANIPULATION SYSTEM

Elbow joint (pitch)

CCTV cameras and lights allow work to be viewed

Wrist joint (yaw, pitch, and roll)

Shoulder joint (yaw and pitch)

End effector grapples satellites and other payloads

Canadarm
The Shuttle cargo bay incorporated a remote-controlled robotic manipulator arm called Canadarm.

Cargo bay doors act as radiators to regulate temperature in space

Pressurized Spacelab module within cargo bay

Canadarm used for grappling satellites and other cargo

Control systems located on flight deck

Middeck contains equipment and living quarters

CARGO BAY DOOR

CARGO BAY

FLIGHT DECK

CENTER FUSELAGE

FORWARD FUSELAGE

DELTA WING

Shuttle is covered in more than 24,000 heat-resistant tiles

SHUTTLE ORBITER

THERMAL PROTECTION SYSTEM

While other spacecraft use heat shields that break away and carry heat with them during reentry, the orbiter's hull was protected by several types of permanent insulation. The ceramic tiles used for the hottest areas proved vulnerable to damage and wear and resulted in one disastrous failure.

TILE

High-temperature reusable surface insulation tile

Heat-resistant borosilicate coating

Heat-absorbing silicon-based foam

Insulated upper surface is relatively cool

TOP

Extremely high temperatures

UNDERSIDE

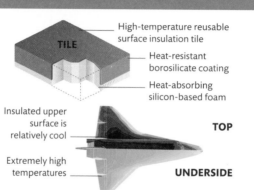

Shuttle rolls upside-down to reduce aerodynamic stress

Main engines and SRBs ignite for launch thrust

The Space Shuttle

NASA's Space Shuttle was a revolutionary launch system that combined conventional rockets with a reusable spaceplane the size of a small airliner. The Shuttle provided the US with access to space from 1981 to 2011.

HOW MANY SPACE SHUTTLES WERE THERE?

NASA's fleet included four flight-worthy orbiters—initially, *Columbia*, *Challenger*, *Discovery*, and *Atlantis* (plus the prototype *Enterprise*). Two Shuttles were lost in accidents, and in 1992, *Endeavour* was built.

Mission profile

The Space Shuttle launched vertically, with the orbiter strapped to a large external fuel tank (ET) that delivered fuel to the orbiter's three main engines. Solid rocket boosters (SRBs) attached to either side of the ET aided the launch. Once in space, the Shuttle used its Orbital Maneuvering System (OMS) to complete operations. After a week or more in space, the orbiter reversed its orientation and fired its main engines to reenter Earth's atmosphere, returning to a horizontal landing as an unpowered glider.

WEIGHING **121 TONS (110 TONNES)** AT LAUNCH, THE **SHUTTLE ORBITER** WAS BY FAR THE **HEAVIEST SPACECRAFT** EVER **PUT INTO ORBIT**

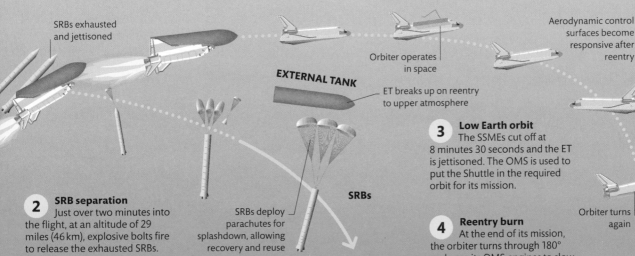

SRBs exhausted and jettisoned

Orbiter operates in space

EXTERNAL TANK

ET breaks up on reentry to upper atmosphere

Aerodynamic control surfaces become responsive after reentry

SRBs deploy parachutes for splashdown, allowing recovery and reuse

SRBs

3 Low Earth orbit
The SSMEs cut off at 8 minutes 30 seconds and the ET is jettisoned. The OMS is used to put the Shuttle in the required orbit for its mission.

Orbiter turns again

2 SRB separation
Just over two minutes into the flight, at an altitude of 29 miles (46 km), explosive bolts fire to release the exhausted SRBs.

4 Reentry burn
At the end of its mission, the orbiter turns through 180° and uses its OMS engines to slow down, reorienting to fly nose first before it enters the atmosphere.

1 Launch
Thrust from all three Space Shuttle Main Engines (SSMEs, fueled from the external tank) plus two SRBs is needed to lift the Shuttle off the ground.

5 Glide approach
The orbiter returns to Earth in a controlled glide, braking from hypersonic speed in a series of computer-controlled turns before the pilots take over for final approach.

Undercarriage deploys 10 seconds before landing

Orbiter begins glide

Space stations

Semipermanent outposts in space increase the time that astronauts can stay in orbit, allowing them to conduct long-duration experiments in zero-gravity and the near-vacuum of space.

The International Space Station

The biggest space station ever built, the International Space Station (ISS) circles Earth in low Earth orbit. Fifteen pressurized modules, including European, US, Russian, and Japanese laboratories, provide living and working space for an average crew of six astronauts. They are connected to the main beam, called a truss structure. On its exterior, the station has multiple robotic arms for various tasks, alongside areas for exposing experiments to space. Power is supplied by tilting solar panels connected to the truss, with a span wider than a soccer field.

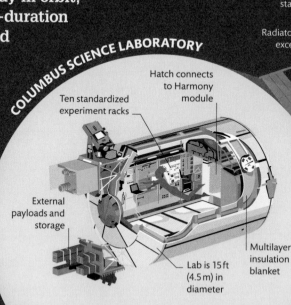

COLUMBUS SCIENCE LABORATORY

Russian Zvezda module includes sleeping quarters for two cosmonauts

Main truss forms station's backbone

Radiators shed excess heat

Hatch connects to Harmony module

Ten standardized experiment racks

External payloads and storage

Lab is 15 ft (4.5 m) in diameter

Multilayer insulation blanket

Tilting, double-sided solar arrays generate power for station

The European Space Agency's Columbus Laboratory was installed by the Space Shuttle *Atlantis* in 2008. It is one of the key ISS laboratories, with experiment space shared between ESA and NASA.

Getting into orbit

Building the ISS was the most complex engineering task ever undertaken in space. Main construction lasted from 1998 to 2011, with the US Space Shuttle playing a crucial role in delivering components and linking them together with its robotic arm. Crews (usually groups of three overlapping each other in six-month expeditions) initially arrived on the Shuttle or Russian Soyuz spacecraft. In 2011, Soyuz became the sole means of access, but commercial space vehicles are now taking some of the burden.

THE INTERNATIONAL SPACE STATION HAS BEEN CONTINUALLY CREWED SINCE OCTOBER 31, 2000

Orbiting the Earth

The ISS orbits at an average altitude of 254 miles (409 km) above Earth, tilted at an angle of 51.6° relative to Earth's equator. This means it circles the Earth once every 92.7 minutes, or 15.5 times every day. The station has an average orbiting speed of 17,227 mph (27,724 kph).

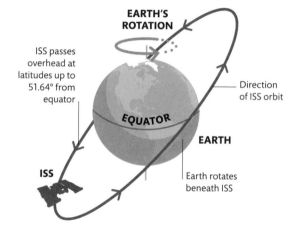

EARTH'S ROTATION

ISS passes overhead at latitudes up to 51.64° from equator

Direction of ISS orbit

EQUATOR

EARTH

ISS

Earth rotates beneath ISS

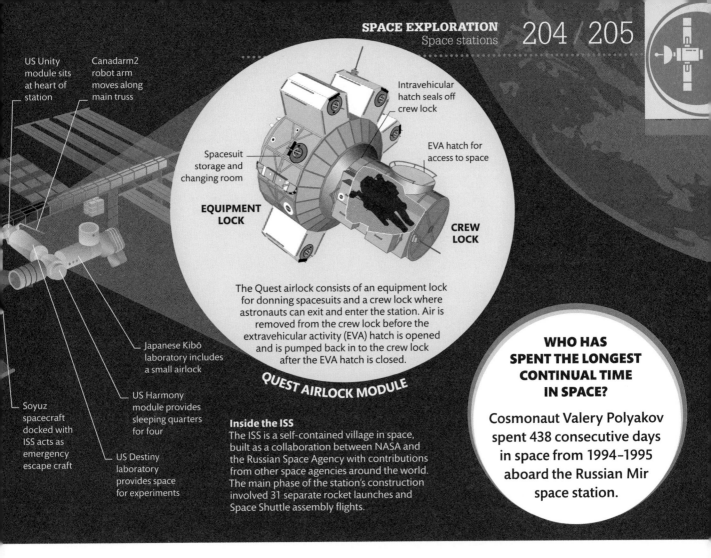

US Unity module sits at heart of station

Canadarm2 robot arm moves along main truss

Intravehicular hatch seals off crew lock

Spacesuit storage and changing room

EVA hatch for access to space

EQUIPMENT LOCK

CREW LOCK

The Quest airlock consists of an equipment lock for donning spacesuits and a crew lock where astronauts can exit and enter the station. Air is removed from the crew lock before the extravehicular activity (EVA) hatch is opened and is pumped back in to the crew lock after the EVA hatch is closed.

QUEST AIRLOCK MODULE

Japanese Kibō laboratory includes a small airlock

Soyuz spacecraft docked with ISS acts as emergency escape craft

US Harmony module provides sleeping quarters for four

US Destiny laboratory provides space for experiments

Inside the ISS
The ISS is a self-contained village in space, built as a collaboration between NASA and the Russian Space Agency with contributions from other space agencies around the world. The main phase of the station's construction involved 31 separate rocket launches and Space Shuttle assembly flights.

WHO HAS SPENT THE LONGEST CONTINUAL TIME IN SPACE?

Cosmonaut Valery Polyakov spent 438 consecutive days in space from 1994–1995 aboard the Russian Mir space station.

Earth-orbiting space stations

The Salyut space stations of the 1970s followed a basic Soviet military design with a single airlock. In 1973, NASA launched a competitor with Skylab, based on leftover Apollo hardware. Salyut 6 (1977) was the first station with two airlocks, allowing crews to visit or swap over without the station being left empty. Mir (1988–2001) was a forerunner to the ISS design, with multiple pressurized units in a modular arrangement.

OTHER EARTH-ORBITING SPACE STATIONS			
Name	**Country**	**Launch date**	**Information**
Salyut 1	USSR	April 1971	The first in a series of space stations based on a design called Almaz, Salyut 1 was abandoned after its first crew died during their return to Earth.
Skylab	USA	May 1973	NASA's Skylab was adapted from a spare Saturn rocket stage and damaged during launch. It was repaired by its first crew and visited by two more in 1973–1974.
Mir	USSR	February 1986	Built over the course of a decade, Mir grew to incorporate seven pressurized modules. In the 1990s, US Space Shuttles docked with the station.
Tiangong-1	China	September 2011	The prototype Chinese space station Tiangong-1 was visited by one automated spacecraft and two crewed Shenzhou missions during two years of operation.

Landing on other worlds

Successfully landing on the surface of another world often requires far more complex systems than just retrorockets—especially so when the atmosphere is substantially thicker or thinner than Earth's.

Curiosity on Mars

The challenges of reaching the Martian surface vary with the size of spacecraft involved. Mars's atmosphere creates substantial friction, so an incoming probe must be shielded from the heat. It is too thin for parachutes alone to slow down the heaviest landers but sufficiently dense to create instability if relying on retrorockets. The Curiosity rover combined a variety of techniques to ensure a safe touchdown.

Landing on Mars

Curiosity's descent combined aerobraking, parachutes, and a complex device called a Sky Crane in an operation that, once triggered, took place with no direct control from Earth.

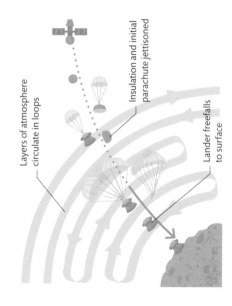

Cruise stage in orbit

1,016 seconds to touchdown

896 seconds to touchdown

Entry into atmosphere

ALTITUDE: 78 MILES (125 KM)

416 seconds to touchdown

Peak heating of probe's heat shielding

1 Mars final approach

The Curiosity rover, safely encased in a two-part aeroshell, separates from the cruise stage in orbit and descends toward the Martian surface.

2 Aerobraking

Friction with the upper atmosphere slows Curiosity from 3.6 miles (5.8 km) per second down to around 1,540 ft (470 m) per second in four minutes.

WHO MADE THE FIRST LANDING ON VENUS?

The Soviet Union's Venera 7 was the first soft-landing probe to reach the surface of Venus intact, sending back data for just 20 minutes.

Landing on Venus

Landing on Venus is even more hazardous than reaching Mars. The atmosphere is thicker and better able to support a parachute but also highly toxic and corrosive. Nevertheless, a series of heavily shielded Venera spacecraft made safe descents in the 1970s and 1980s.

Insulation and initial parachute jettisoned

Layers of atmosphere circulate in loops

Lander freefalls to surface

A hazardous descent

Venera landers used a combination of aerobraking and parachutes to reach the surface of Venus. The thick atmosphere cushioned the final 30-mile (50-km) fall.

BOUNCING DOWN ON MARS

In 2004, a pair of rovers arrived on Mars by using a combination of aerobraking, parachutes, and retrorockets, finally dropping to the surface encased in airbags.

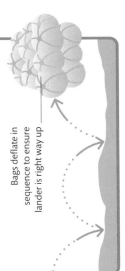

Bags deflate in sequence to ensure lander is right way up

SKY CRANE DESCENT

The Sky Crane system lowered Curiosity to a gentle, soft landing on the surface of Mars before flying away.

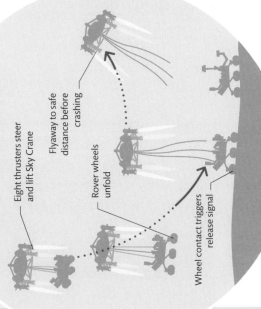

Eight thrusters steer and lift Sky Crane

Flyaway to safe distance before crashing

Rover wheels unfold

Wheel contact triggers release signal

Heat shield separates and radar begins data collection

ALTITUDE: 6 MILES (10 KM)

ALTITUDE: 7 MILES (11 KM)

Parachute diameter 52ft (16m)

162 seconds to touchdown

3 Parachutes

A parachute deploys at supersonic speed and unfolds, slowing the probe's descent to about 330ft (100m) per second.

138 seconds to touchdown

Powered descent

ALTITUDE: 1.1 MILES (1.8 KM)

4 Sky Crane

In the final descent phase, the rover is transported to its landing site beneath a flying platform called the Sky Crane.

Rover lowered beneath Sky Crane from 65ft (20m)

CURIOSITY ENTERED THE MARTIAN ATMOSPHERE AT A SPEED OF 3.6 MILES (5.8 KM) PER SECOND

Mars rovers

Until humans can safely explore other planets, wheeled mobile robots, called rovers, are the next best alternative. So far, humans have sent five rovers to the planet Mars, each one more sophisticated and capable of answering more complex scientific questions than the last.

SOJOURNER
Length: 2 ft (65 cm)

SPIRIT AND OPPORTUNITY
Length: 5 ft (1.6 m)

CURIOSITY
Length: 10 ft (3 m)

PERSEVERANCE
Length: 6.5 ft (2 m)

Rover sizes
Rovers vary in size and complexity depending on the objectives of their missions on Mars.

WHAT WAS THE FIRST ROVER ON ANOTHER WORLD?

The USSR's Lunokhod 1 was a solar-powered vehicle that landed on the Moon in November 1970. It operated for almost 10 months.

The Curiosity Mars Rover

In 2012, the car-sized Curiosity rover landed in an ancient lake bed where scientists hoped to find evidence for hospitable conditions in the Martian past. The most advanced rover ever to land on Mars, it carried onboard laboratories, advanced cameras, weather instruments, and a versatile arm for drilling rocks and collecting samples.

Analyzing the Martian surface
Curiosity is equipped with many scientific instruments, including a laser-powered spectrometer capable of identifying rock samples at a distance.

Multiple cameras for navigation and analysis

MastCam takes high-resolution color images

ChemCam utilizes laser, with range of up to 23 ft (7 m), to vaporize rock layers and soil

MAST

LASER

Electricity generated by heat from decay of radioactive plutonium stored inside power unit

POWER UNIT HOUSING

Ultra-High Frequency (UHF) antenna for communication with orbiting satellites

UHF ANTENNA

Sensors monitor wind speed, wind direction, and air temperature

WEATHER STATION

HIGH-GAIN ANTENNA

RADIATION DETECTOR

Arm tools include camera, drill, and X-ray spectrometer

NEUTRON SPECTROMETER

DRILL

ROBOTIC ARM

Drill extracts samples

Wheels can surmount obstacles up to 26 in (65 cm) high

Mars Descent Imager camera

INTERNAL LABORATORIES

Arm is 6.5 ft (2 m) long

Driving on Mars

In order to navigate the uneven surface of Mars, rovers are equipped with a rocker-bogie suspension system to keep level. The delay in sending radio signals back and forth to Earth means that engineers cannot steer the vehicle in real time—instead they gather data and images before planning a course to each new waypoint. The rover then follows this route, using sensors and its onboard computer to navigate minor hazards along the way.

CURIOSITY HAS A **TOP SPEED** OF JUST **295 FT (90 M) PER HOUR**

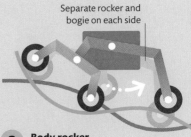

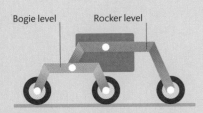

Bogie level   Rocker level

1 **Wheels for Mars**
Curiosity drives on six large wheels made of aluminum, with treads to grip the rocky surface. Each wheel has an independent drive motor, and the front and rear wheels have steering motors.

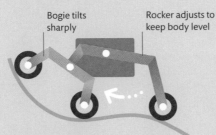

Bogie tilts sharply

Rocker adjusts to keep body level

2 **Rear bogie**
On each side of the rover, its center and rear wheels are connected to a frame, known as a bogie, that can tilt to keep both wheels in roughly equal contact with the Martian terrain.

Separate rocker and bogie on each side

3 **Body rocker**
The bogie and front wheel on each side attach to the rover body through a larger pivoting frame, the rocker. This means that all six wheels can be at different levels without throwing the rover off balance.

Overhead view
Curiosity's six-wheel drive, with no connecting axles between the two sides, allows the rover to function even if some wheels get stuck in sand or damaged and disabled by sharp rocks.

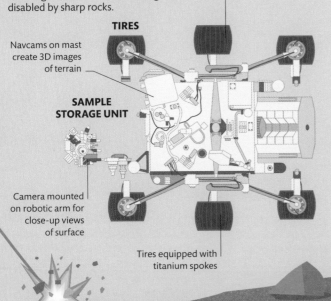

Tires have tread made of 24 chevrons

TIRES

Navcams on mast create 3D images of terrain

SAMPLE STORAGE UNIT

Camera mounted on robotic arm for close-up views of surface

Tires equipped with titanium spokes

OTHER ROVERS ON MARS

The first rover to land on Mars, in 1997, was the small solar-powered Sojourner, part of the Mars Pathfinder mission. This was followed by the larger Mars Exploration Rovers Spirit and Opportunity in 2004; Curiosity (the Mars Science Laboratory) in 2012; and Perseverance, launched in 2020.

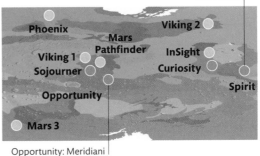

Spirit: Gusev Crater, 2004–2010

Phoenix

Viking 2

Mars Pathfinder

Viking 1
Sojourner

InSight

Curiosity

Opportunity

Spirit

Mars 3

Opportunity: Meridiani Planum, 2004–2018

⬤ Other landers   ⬤ Mars rovers

GRAVITATIONAL SLINGSHOTS

The Voyagers relied on a technique called the gravity assist or slingshot. This allows a spacecraft to alter its direction and speed without an engine burn by falling into the gravitational field of a moving planet at just the right angle. From the point of view of the planet, the spacecraft approaches and leaves at the same speed, but relative to the Sun and wider Solar System, its speed is altered.

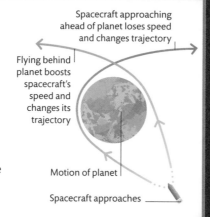

Spacecraft approaching ahead of planet loses speed and changes trajectory

Flying behind planet boosts spacecraft's speed and changes its trajectory

Motion of planet

Spacecraft approaches

WHY DID VOYAGER 1 NOT GO TO URANUS AND NEPTUNE?

NASA scientists wanted at least one Voyager spacecraft to investigate Saturn's giant moon Titan. This required an approach trajectory that aimed below Saturn's south pole, which deflected the spacecraft out of the plane of the Solar System.

Planetary alignment

The Voyager missions were made possible by a grand alignment of all four outer planets in the late 1970s that saw Jupiter, Saturn, Uranus, and Neptune arranged along a spiral trajectory. This alignment, which only happens every 175 years, made it possible for each spacecraft to fly past each planet in turn without using huge amounts of fuel to alter their flight path.

ORBIT OF NEPTUNE

ORBIT OF URANUS

Grand tours

Launched in 1977, the two Voyager spacecraft provided humanity with a first detailed look at the giant planets of the outer Solar System. They continue to send back valuable scientific data even today.

Interstellar missions

Although the Voyager spacecraft are now well beyond the orbits of the planets, they are still sending back valuable information about conditions at the edge of the Solar System. This is where the heliosphere—the region filled by the solar wind of particles flowing out from the Sun at high speeds—merges into interstellar space. Both probes will continue to transmit data until their electricity supplies run out in the mid-2020s.

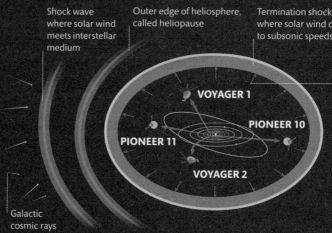

Shock wave where solar wind meets interstellar medium

Outer edge of heliosphere, called heliopause

Termination shock—where solar wind drops to subsonic speeds

Outward flow of solar wind

VOYAGER 1

PIONEER 10

PIONEER 11

VOYAGER 2

Travel beyond the Solar System
The Voyagers are not the only probes leaving the Solar System. They are accompanied by Pioneers 10 and 11, which flew past Jupiter and Saturn, and New Horizons.

Galactic cosmic rays

The Voyager spacecraft

Each Voyager probe was built around a decahedral (10-sided) that held the main spacecraft systems and most of the scientific instruments. Long antennae emerging from the body measured magnetic fields and radio waves, while an antenna dish allowed communication with Earth. A steerable platform on the end of a boom allowed cameras and some other instruments to keep planets and moons in view.

Spectrometers measure thermal, structural, and compositional nature of targets

VOYAGER 2

Golden record contains collection of data about Earth and is carried on each spacecraft

Radiators for shedding excess heat

12.1-ft (3.7-m) high-gain antenna

ANTENNA

Hydrazine thruster

Radioisotope thermal generator (electricity source) on boom to avoid interference with instruments

Voyager 1 performs Saturn flyby and encounters Titan on November 12, 1980

Flyby of Jupiter on March 5, 1979

Voyager 1 launches from Earth on September 5, 1977

Voyager 2 launches from Earth on August 20, 1977

Earth

Flyby of Jupiter on July 9, 1979

Low-field magnetometer boom

Saturn

Voyager 2 performs Saturn flyby on August 26, 1981

Uranus

Voyager 2 performs flyby of Uranus on January 24, 1986

Neptune flyby on August 25, 1989

Voyager's tools

Alongside a magnetometer and radio antenna, Voyager's main instruments included an imaging camera, spectrometers to analyze the chemistry of planetary atmospheres, and instruments to detect particles in interplanetary space.

Neptune

Due to Titan encounter, Voyager 1 is unable to perform subsequent flybys

Planetary pinball

After launching from Earth, the two Voyager spacecraft flew past first Jupiter and then Saturn. Voyager 2 continued to Uranus and Neptune, while Voyager 1 was deflected onto a path that took it out of the plane of the Solar System.

VOYAGER 1

VOYAGER 1 OFFICIALLY BECAME THE **FIRST ARTIFICIAL OBJECT** TO ENTER INTERSTELLAR SPACE ON **AUGUST 25, 2012**

Voyage to Saturn

Putting a spacecraft in orbit around a planet requires a very different trajectory from a simple flyby. In order to approach Saturn at the correct angle, Cassini followed a seven-year flight path involving several gravity-assist maneuvers.

1 Venus assists
In 1998 and 1999, Cassini made two flybys of Venus. The first boosted its speed by 4 miles (7 km) per second, but it had to be slowed with an engine burn to put it on course for a second flyby and speed boost.

2 Return to Earth
In August 1999, Cassini flew past Earth at an altitude of 728 miles (1,171 km). The spacecraft gained another speed boost of 3.4 miles (5.5 km) per second, putting it on course for a flyby of Jupiter.

Second flyby of Venus

First flyby of Venus

EARTH'S ORBIT

SUN

Spacecraft engages in a Venus targeting maneuver

Launch of spacecraft

Cassini flyby of Earth

JUPITER'S ORBIT

Jupiter flyby increases Cassini's velocity

Orbiting Saturn
During Cassini's 13 years at Saturn, its orbit was repeatedly changed using gravity assists (mostly from Titan) and occasional engine burns to ensure close encounters with the planet's many moons.

Path from Earth

HUYGENS-TITAN ENCOUNTER

SATURN

Fourth orbit

Third orbit

Second orbit

Orbit of Titan

First orbit

Orbit of Iapetus

Spacecraft reaches Saturn's orbit

4 Arrival at Saturn
In mid-2004, Cassini successfully entered the Saturn system and used its main engine in two maneuvers that shed speed and dropped it into an initial elliptical orbit of the planet.

3 Jupiter swingby
In December 2000, Cassini flew past Jupiter at a distance of 6 million miles (9.7 million km). It conducted observations of the Solar System's largest planet and received a further boost to its speed.

Mass spectrometer for analyzing captured particles

Low-gain antenna

Huygens probe before deployment to Titan

High-gain antenna

Mapping cameras and spectrometers

Cassini's instruments
Cassini carried a variety of instruments. Radar allowed it to pierce Titan's atmosphere, while visible, infrared, and ultraviolet cameras captured a huge variety of information.

Dual main rocket engines

The Cassini orbiter

The bus-sized Cassini spacecraft remains the most complex uncrewed spacecraft sent into space by NASA. Launched in 1997, it orbited Saturn between 2004 and 2017, sending back a wealth of information about the planet, its rings, and its huge family of moons. The spacecraft also carried Huygens, a Titan lander built by the European Space Agency (ESA) that was released five months after Cassini's arrival in orbit. At the end of its mission, Cassini was crashed into Saturn's atmosphere to avoid possible contamination of its moons.

HUYGENS ON TITAN

The Huygens lander carried a variety of scientific instruments to investigate conditions on Titan. Uniquely, the probe was designed to float, because extensive lakes of liquid hydrocarbon chemicals were expected on Titan's surface.

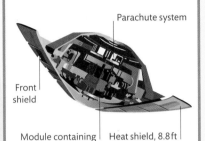

Parachute system

Front shield

Module containing scientific instruments

Heat shield, 8.8 ft (2.7 m) in diameter

HUYGENS PROBE

GASES AROUND THE **GALILEO PROBE** REACHED TEMPERATURES OF 28,000°F (15,500°C), BURNING AWAY ITS HEAT SHIELD

HOW BIG WAS CASSINI?

The Cassini spacecraft was 22.3 ft (6.8 m) long and 13 ft (4 m) wide, with a mass of 4,740 lb (2,150 kg), plus 6,905 lb (3,132 kg) of rocket propellant.

Orbiting giants

The Grand Tour flybys of the 1980s (see pp.210–211) were followed by more detailed explorations of the giant planets Jupiter and Saturn using complex spacecraft that remained in orbit for years.

The Galileo mission

The Galileo spacecraft orbited Jupiter from 1995 to 2003 and successfully carried out multiple flybys of the planet and its four giant satellites: Io, Europa, Ganymede, and Callisto (see pp.68–71). Galileo shed its excess speed without a retrorocket burn thanks to a daring aerobraking strategy in which it slowed down by dipping into the upper layers of Jupiter's atmosphere. Shortly after its arrival, the spacecraft deployed an atmospheric probe that parachuted into Jupiter's clouds and sent back valuable data about their composition.

Probing Jupiter's atmosphere

Galileo's atmospheric probe entered Jupiter's gassy outer layers at a speed of around 30 miles (48 km) per second. In the space of two minutes, the probe slowed to subsonic speeds before deploying its parachute.

Probe enters Jupiter's atmosphere

Main parachute is deployed but buffeted by winds of up to 380 mph (610 kph)

Probe passes through cloud layer of small condensed particles

WIND

Probe's heat shield detaches during descent

CLOUD LAYER

Radio contact ceases after 78 minutes due to high heat in Jupiter's atmosphere

INTERIOR OF JUPITER

HOW CLOSE DID NEW HORIZONS COME TO PLUTO?

The craft flew 7,770 miles (12,500 km) above Pluto's surface—passing through the orbits of the dwarf planet's five moons like an arrow hitting a bull's-eye.

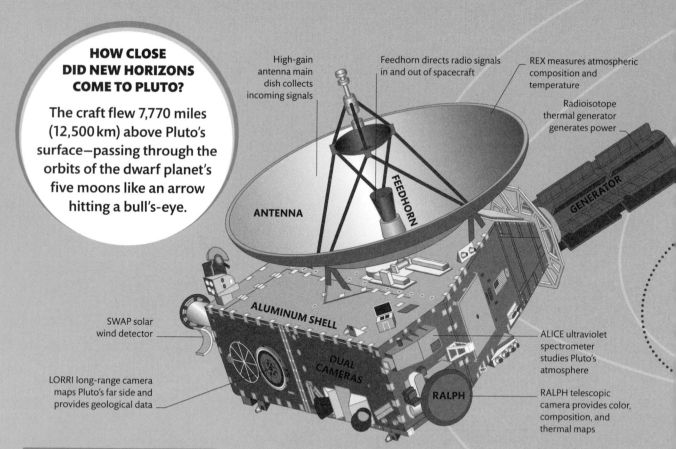

High-gain antenna main dish collects incoming signals

Feedhorn directs radio signals in and out of spacecraft

REX measures atmospheric composition and temperature

Radioisotope thermal generator generates power

FEEDHORN

ANTENNA

GENERATOR

ALUMINUM SHELL

SWAP solar wind detector

LORRI long-range camera maps Pluto's far side and provides geological data

DUAL CAMERAS

RALPH

ALICE ultraviolet spectrometer studies Pluto's atmosphere

RALPH telescopic camera provides color, composition, and thermal maps

NEW HORIZONS' FLYBY

Following the encounter with Pluto, NASA was eager to send New Horizons to another Kuiper Belt object. Dwindling fuel limited their choices, but with a minor adjustment to the spacecraft's flight path, New Horizons was able to fly past and take images of a small world called Arrokoth on January 1, 2019.

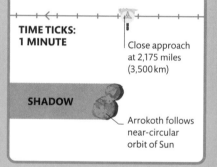

TIME TICKS: 1 MINUTE

Close approach at 2,175 miles (3,500 km)

SHADOW

Arrokoth follows near-circular orbit of Sun

Packing for Pluto

With the entire mass of New Horizons limited to 884 lb (401 kg) plus propellant for its thrusters, the spacecraft had room for just 67 lb (30 kg) of instruments. Power was also an issue, since the amount of fuel that could be carried to generate electricity was limited. Fortunately, scientists and engineers benefited from advances in microelectronics and were able to pack in seven separate instruments that operated on less than 28 watts in total.

The path to Pluto
After leaving Earth, New Horizons flew past Jupiter a year later, receiving a gravity assist that boosted its speed. It then entered hibernation mode until late 2014, when it was awoken in preparation for the Pluto encounter.

Transmitting data
Sending radio signals from the edge of the Solar System is a challenge. With bandwidth needed during encounters for critical commands and navigation, New Horizons recorded its science data onto solid-state recorders, then sent it back to Earth over several months.

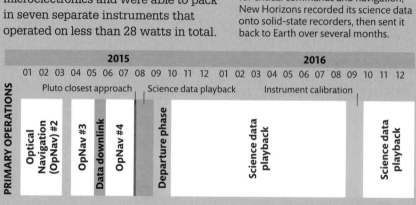

	2015											2016											
	01	02	03	04	05	06	07	08	09	10	11	12	01	02	03	04	05	06	07	08	09	10	11	12

Pluto closest approach   Science data playback   Instrument calibration

PRIMARY OPERATIONS

Optical Navigation (OpNav) #2

OpNav #3

Data downlink

OpNav #4

Departure phase

Science data playback

Science data playback

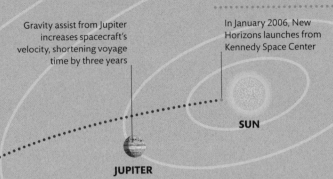

Gravity assist from Jupiter increases spacecraft's velocity, shortening voyage time by three years

In January 2006, New Horizons launches from Kennedy Space Center

SUN

JUPITER

PLUTO ENCOUNTER

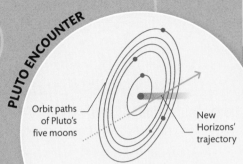

Orbit paths of Pluto's five moons

New Horizons' trajectory

New Horizons flew past Pluto at a speed of more than 52,200 mph (84,000 kph). The spacecraft took close-up images of Pluto, studied its atmosphere, and measured its mass.

Racing to Pluto

Despite no longer being classified as a major planet, Pluto is still among the largest objects in the Kuiper Belt (see pp.82–83) at the edge of our Solar System. In 2006, NASA launched New Horizons, a spacecraft that aimed to reach this dwarf planet while it remained relatively close to the Sun.

PLUTO

Flyby of Pluto in July 2015

Planning the mission

Pluto's elongated orbit means that its distance (and the ease of reaching it from Earth) varies significantly. Furthermore, conditions on the dwarf planet's surface were expected to change considerably depending on the amount of light reaching it from the Sun. Because Pluto was retreating from its closest approach to the Sun, which happened in 1989, time was of the essence— it was vital to keep the spacecraft light and fast.

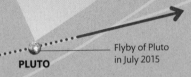

NEW HORIZONS WAS THE **FASTEST SPACECRAFT** EVER LAUNCHED FROM EARTH, **LEAVING ORBIT AT A SPEED OF 10 MILES (16 KM) PER SECOND**

Launch boost

In order to send New Horizons on its way at high-enough speed, the spacecraft was launched with a unique rocket configuration—a powerful Atlas 5b two-stage rocket assisted by an unprecedented five solid rocket boosters clustered at the base and topped by a Star 48B third stage. This allowed the rocket to achieve, within just 45 minutes of launch, the speed necessary to escape the Solar System.

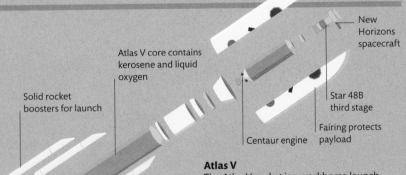

New Horizons spacecraft

Atlas V core contains kerosene and liquid oxygen

Solid rocket boosters for launch

Star 48B third stage

Centaur engine

Fairing protects payload

Atlas V
The Atlas V rocket is a workhorse launch vehicle that typically consists of the Atlas V first stage and a Centaur second stage, supported by a number of solid rocket boosters at the base.

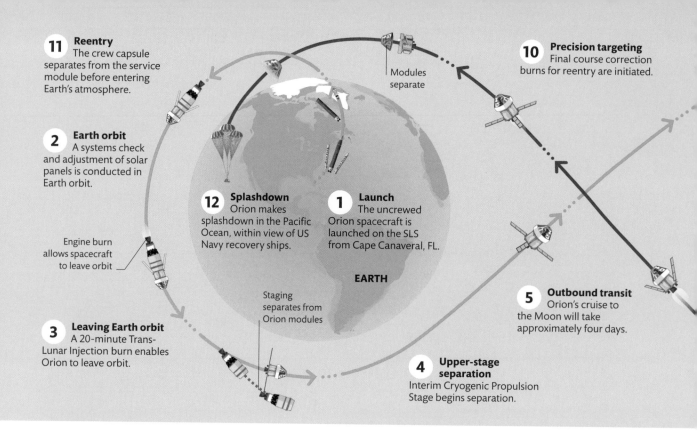

11 Reentry
The crew capsule separates from the service module before entering Earth's atmosphere.

2 Earth orbit
A systems check and adjustment of solar panels is conducted in Earth orbit.

Engine burn allows spacecraft to leave orbit

3 Leaving Earth orbit
A 20-minute Trans-Lunar Injection burn enables Orion to leave orbit.

Modules separate

12 Splashdown
Orion makes splashdown in the Pacific Ocean, within view of US Navy recovery ships.

1 Launch
The uncrewed Orion spacecraft is launched on the SLS from Cape Canaveral, FL.

EARTH

Staging separates from Orion modules

4 Upper-stage separation
Interim Cryogenic Propulsion Stage begins separation.

10 Precision targeting
Final course correction burns for reentry are initiated.

5 Outbound transit
Orion's cruise to the Moon will take approximately four days.

Future spacecraft

In the near future, astronauts will travel in a variety of spacecraft, from commercial ferries running to and from the International Space Station (ISS), through suborbital capsules for space tourism, to advanced vehicles designed to explore the wider Solar System.

THE SLS BLOCK 2 VARIANT WILL LAUNCH 143 TONS (130 TONNES) TO EARTH ORBIT

The Orion MPCV

The Orion Multi-Purpose Crew Vehicle (MPCV) is a versatile new spacecraft designed by NASA for a variety of new exploration missions. Looking a little like an outsized Apollo spacecraft, it is designed to carry four to six astronauts on missions of up to 21 days without support. Orion will be launched atop NASA's new Space Launch System (SLS)—a multipurpose rocket that can also put components of larger spacecraft intended for long-duration interplanetary exploration into orbit.

Saturn successor
Initially derived from tried and tested elements of NASA's Space Shuttle program, the SLS can be configured in various blocks, the most powerful of which can put 20 percent more payload into orbit than the Saturn V rocket.

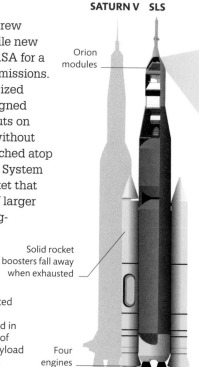

SATURN V   SLS

Orion modules

Solid rocket boosters fall away when exhausted

Four engines

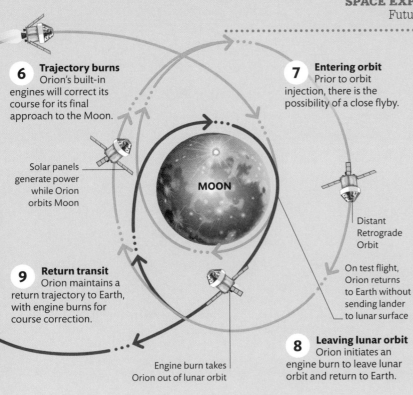

6 **Trajectory burns**
Orion's built-in engines will correct its course for its final approach to the Moon.

Solar panels generate power while Orion orbits Moon

9 **Return transit**
Orion maintains a return trajectory to Earth, with engine burns for course correction.

Engine burn takes Orion out of lunar orbit

MOON

7 **Entering orbit**
Prior to orbit injection, there is the possibility of a close flyby.

Distant Retrograde Orbit

On test flight, Orion returns to Earth without sending lander to lunar surface

8 **Leaving lunar orbit**
Orion initiates an engine burn to leave lunar orbit and return to Earth.

The future of Moon exploration

Orion and the SLS form the backbone of NASA's Artemis program—an ambitious plan for a return to the Moon by around 2024. The program involves establishment of a lunar gateway space station in orbit around the Moon, new cargo ferries to deliver supplies, and a new Human Landing System spacecraft to put men and women on the surface at the lunar south pole and support them for up to a week.

First step to the Moon
The initial Artemis 1 mission is an uncrewed flight to test key components of the SLS and Orion in Earth and lunar orbit.

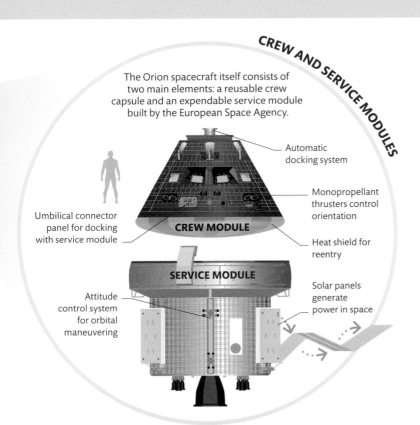

CREW AND SERVICE MODULES

The Orion spacecraft itself consists of two main elements: a reusable crew capsule and an expendable service module built by the European Space Agency.

Automatic docking system

Monopropellant thrusters control orientation

CREW MODULE

Umbilical connector panel for docking with service module

Heat shield for reentry

SERVICE MODULE

Attitude control system for orbital maneuvering

Solar panels generate power in space

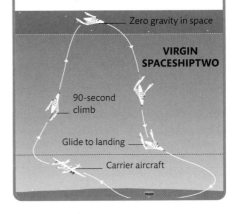

SPACE TOURISM

The next decade will see various companies offering space tourism. Virgin Galactic's entry is the revolutionary SpaceShipTwo, a reusable, shuttlelike capsule that launches from a high-altitude carrier aircraft and powers to the edge of space using rockets before drifting back to Earth.

Zero gravity in space

VIRGIN SPACESHIPTWO

90-second climb

Glide to landing

Carrier aircraft

Index

Acknowledgments

DK would like to thank the following people for help in preparing this book: Giles Sparrow for help with planning the contents list; Helen Peters for compiling the index; Katie John for proofreading; Senior DTP Designer Harish Aggarwal; Jackets Editorial Coordinator Priyanka Sharma; and Managing Jackets Editor Saloni Singh.